Morality in a Technologic…

The technological advances of contemporary society have outpaced our moral understanding of the problems that they create. How will we deal with profound ecological changes, human cloning, hybrid people, and eroding cyberprivacy, just to name a few issues? In this book, Lorenzo Magnani argues that existing moral constructs often cannot be applied to new technology. He proposes an entirely new ethical approach, one that blends epistemology with cognitive science. The resulting moral strategy promises new dignity for overlooked populations, both of today and of the future.

Lorenzo Magnani, philosopher and cognitive scientist, is a professor at the University of Pavia, Italy, and the director of its Computational Philosophy Laboratory. He is the author of *Abduction, Reason, and Science* and in 1998 started the series of International Conferences on Model-Based Reasoning.

Morality in a Technological World

Knowledge as Duty

LORENZO MAGNANI
University of Pavia

CAMBRIDGE
UNIVERSITY PRESS

CAMBRIDGE UNIVERSITY PRESS
Cambridge, New York, Melbourne, Madrid, Cape Town, Singapore,
São Paulo, Delhi, Dubai, Tokyo

Cambridge University Press
32 Avenue of the Americas, New York, NY 10013-2473, USA

www.cambridge.org
Information on this title: www.cambridge.org/9780521121798

First published 2007
This digitally printed version 2009

A catalog record for this publication is available from the British Library

Library of Congress Cataloging in Publication data

Magnani, Lorenzo.
Morality in a technological world : knowledge as duty / Lorenzo Magnani.
p. cm.
Includes bibliographical references and index.
ISBN 978-0-521-87769-5 (hardback)
1. Technology – Moral and ethical aspects. 2. Ethics, Modern – 21st century. I. Title.
BJ59.M34 2007
170 – dc22 3 2006100081

ISBN 978-0-521-87769-5 Hardback
ISBN 978-0-521-12179-8 Paperback

To my wife, Anna

Contents

Preface

RESPECTING PEOPLE AS THINGS

When my institution, the University of Pavia, was founded near Milan in 1361, women were viewed very differently from the way we see women today. Back then, a woman living in one of the centuries-old houses I pass each day on my way to work here in Pavia would not have been considered as "human" as a man: seven hundred years ago, she would have essentially been property – first her father's and then, later, her husband's – and she would probably have had little control over matters concerning her family or her own destiny. There is a very good chance that she would have been illiterate, as were most women and many men in medieval Europe, and she certainly would not have been permitted to attend the city's then-new university. In the fourteenth century, the centers of learning in northern Italy were among the most advanced in the world, yet even they considered women to be unworthy or incapable of being educated. Indeed, it took nearly 400 years for women to gain admission to the University of Pavia, and not until the eighteenth century was a degree awarded to a woman: Maria Pellegrini Amoretti (who was from a wealthy family, not surprisingly) took a law degree in 1777.

At the institution today, however, women work alongside men as both students and faculty members, and while we must continue to strive for gender equity, the current level of intellectual interaction between female and male scholars would have been unimaginable in the medieval world. Attitudes toward gender roles did not evolve because of some inherent change in women, of course, but because people have learned a great deal more about the human condition since 1361. By what mechanism did this shift occur? What knowledge allowed humankind to change the way it views women? And for the purposes of this book, how can we learn from that transformation so that others may enjoy greater status as well?

Before turning to this book's central theme of regarding knowledge as duty, it is useful to think about how knowledge can affect an entity's moral status. In addition to women, many other kinds of entities – both living and nonliving – that were once considered much less valuable than they are today have also acquired a different kind of moral worth: *intrinsic* value, or value as an end in itself. An entity's intrinsic value, of course, arises not from a change in the thing itself but from changes in human thinking and knowledge; if various acts of cognition can imbue things with new moral value, I submit that certain undervalued human beings can reclaim the sort of moral esteem currently held by some "external things," like endangered species, artworks, databases, and even some overvalued political institutions.

As the subtitle *Knowledge as Duty* suggests, morality is distributed in our technological world in a way that makes some scientific problems particularly relevant to ethics: ecological imbalances, the medicalization of life, and advances in biotechnology – themselves all products of knowledge – seem to me to be especially pertinent topics of discussion. The system of designating certain animals as endangered, for example, teaches us that there is a continuous delegation of moral values to externalities; this may also cause some people to complain that wildlife receives greater moral and legal protection than, for example, disappearing cultural traditions. I wondered what reasoning process would result in a nonhuman thing's being valued over a living, breathing person and asked myself what might be done to elevate the status of human beings. One solution, I believe, is to reexamine the respect we have developed for particular externalities and then to use those things as a vehicle to return value to people.

The well-known Kantian tradition in ethics teaches that human beings should not be treated solely as "means" or "things" in a merely instrumental way but should, instead, be regarded as "ends." I believe, however, that if we rigidly adhere to Kant's directive, we make it impossible to embrace an important new strategy I propose in Chapter 1: "respecting people as things," the notion that people must be regarded as "means" (things) insofar as these means involve "ends." In essence, the idea holds that human beings often can and even *should* be treated as "things," and that in the process they become "respected as things" that had been ascribed more value than some people. We must reappropriate the instrumental and moral values that people have lavished on external things and objects, which I contend is central to reconfiguring human dignity in our technological world.

The potential benefits of "respecting people as things," then, undermine Kant's traditional distinction between intrinsic value and instrumental value, and they are not the only factors to do so: in Chapter 3, I argue that more advanced and more pervasive technology has also blurred the line between humans and things – machines, for example – and between natural things and artifacts, and that it has become increasingly difficult to discern where the human body ends and the non-human thing begins. We are in a sense "folded into" nonhumans, so that we delegate action to external things (objects, tools, artifacts) that in turn share our human existence with us. It is just this *hybridization* that necessitates treating people as things and, fortunately, that makes this course of action easier to pursue. Again, my counterintuitive conclusion is that instead of treating people as means, we can improve their lives by recognizing their part-thingness and respecting them as things.

In turn, the concept of "respecting people as things" provides an ethical framework through which to analyze the condition of modern people, who, as increasingly commodified beings, are becoming more and more thing-like anyway. In this book, I will use this construct to interrogate the medicalization of life (Chapter 2), cybernetic factors (Chapter 4), and the influences of globalization (Chapter 5).

MORAL MEDIATORS

I have said that only human acts of cognition can add worth to or subtract value from an entity, and that revealing the similarities between people and things can help us to attribute to human beings the kind of worth that is now held by many highly valued nonhuman things. This process suggests a new perspective on ethical thinking: indeed, these objects and structures can mediate moral ideas and recalibrate the value of human beings by playing the role of what I call *moral mediators*.

What exactly is a moral mediator? As I explain in Chapter 6, I derived the concept of the moral mediator from that of the epistemic mediator, which I introduced in my previous research on abduction and creative and explanatory reasoning. First of all, moral mediators can extend value from already-prized things to human beings, as well as to other nonhuman things and even to "non-things" like future people and animals. We are surrounded by human-made and artificial entities, whether they are concrete objects like a hammer or a PC or abstractions like an institution or society; all of these things have the potential to serve as moral mediators. For this reason, I say that it is critically important for current ethics to address not only the relationships among human beings, but also those between human and nonhuman entities. Moreover, by exploiting the concepts of "thinking through doing" and of manipulative abduction, we can see that a considerable part of moral action is performed in a tacit way, so to say, "through doing." Part of this "doing" can be considered a manipulation of the external world in order to build various moral mediators that function as enormous new sources of ethical information and knowledge. I call these schemes of action "templates of moral doing."

In the cases just mentioned, moral mediators are purposefully constructed to achieve particular ethical effects, but other aspects and cognitive roles of moral mediators are equally important: moral mediators are also beings, entities, objects, and structures that objectively, even beyond human beings' intentions, carry possible ethical or unethical consequences.

External moral mediators function as components of a memory system that crosses the boundary between person and environment. For instance, when a society moves an abused child into a foster home, an example I use in Chapter 6, it is seeking both to protect her and to reconfigure her social relationships; in this case, the new setting functions as a moral mediator that changes how she relates to the world – it can supply her with new emotions that bring positive moral and psychological effects and help her gain new perspectives on her past abuse and on adults in general. In *Morality in a Technological World,* I depict these processes as "model-based" inferences, and indeed one way moral mediators transform moral tasks is by promoting further moral inferences in agents at the level of model-based abduction, a concept I introduced in a previous book on abductive reasoning. I use the term "model-based reasoning" to mean the constructing and manipulating of certain representations, not mainly sentential and/or formal, but mental and/or related to external mediators: obvious examples of model-based inferences include building and using visual representations, conducting thought experiments, and engaging in analogical reasoning. In this light, an emotional feeling also can be interpreted as a kind of model-based cognition. Of course, abductive reasoning – the process of reasoning to hypotheses – can be performed in a model-based way, either internally or with the help of external mediators.

Moreover, I can use manipulation to alter my bodily experience of pain; I can, for example, follow the behavior template "control of sense data" described in Chapter 6, during which I might shift – often unconsciously – the position of my body. Through manipulation I can also change my body's relationships with other humans and nonhumans experiencing distress, as did Mother Theresa, whose rich, personal moral feeling and consideration of pain were certainly shaped by her physical proximity to starving and miserable people and by her manipulation of their bodies. In many people, moral training is often related to the spontaneous (and sometimes fortuitous) manipulation of both sense data and their own bodies, for these actions can build morality immediately and nonreflectively "through doing."

Artifacts serve as moral mediators in many situations, as in the case of certain machines that affect privacy. Chapter 4 addresses the fact that the internet mediates human interaction in a much more profound way than do traditional forms of communication like paper, the telephone, and mass media, even going so far as to record interactions in many situations. The problem is that because the internet mediates human identity, it has the power to affect human freedom. Thanks to the internet, our identities today largely consist of an externally stored quantity of data, information, images, and texts that concern us as individuals, and the result is a "cyborg" of both flesh and electronic data that identifies us. I contend that this complex new "information being" depicts new ontologies that in turn involve new moral problems. We can no longer apply old moral rules and old-fashioned arguments to beings that are simultaneously biological and virtual, situated in a three-dimensional local space and yet "globally omnipresent" as information packets. Our cybernetic locations are no longer simple to define, and increasing telepresence technologies will exacerbate this effect, giving external, nonbiological resources even greater powers to mediate ethical endowments such as those related to our sense of who and what we are and what we can do. These and other effects of the internet – almost all of which were unanticipated – are powerful motivators of our duty to construct new knowledge.

I believe that in the context of this abstract but ubiquitous technological presence, certain moral approaches that ethics has traditionally tended to disparage are worth a new look. Taking care of both people and external things through personal, particular acts – a moral orientation often associated with women – rather than relating to others through an impersonal, general concern about humanity has a new appeal. The ethics of care does not consider the abstract "obligation" as essential; moreover, because it does not require that we impartially promote the interests of everyone alike, it allows us to focus on those who most need assistance.

In short, a considerable part of morality occurs in an implicit way, so to say, "through doing," and part of this "doing" features manipulating the external world in order to build various external "moral mediators" that can provide vast amounts of new information and knowledge, transform ethical features and effects, and sometimes, of course, generate unethical outcomes.

MORAL REASONING

In this book, I will consider numerous ethical issues related to technology: ecology, biotechnology, the hybridization of human beings, cyberprivacy, bad faith, globalization, and the unethical effects of external systems and technologies in general. Each of these discussions underscores the importance of producing and exploiting appropriate ethical knowledge and reinforces my argument that knowledge is a duty. If, as I contend, new ethical, scientific, and other kinds of understandings must be developed and implemented, then cognitive concerns also become fundamentally important. In Chapters 6 and 7, I closely examine the cognitive aspects of moral mediators and of other methodological problems related to ethical reasoning and moral deliberation.[1]

Not only are ethical knowledge and reasoning expressed at the verbal/propositional level, they can also involve model-based (visual, for example) and manipulative/"through doing" aspects: for example, an important component of ethics is imagination, which is – together with analogy, visualization, simulation, the thought experiment, and so on – a form of model-based reasoning. Creativity is also important, for through it human beings expand knowledge and create new perspectives. To explain morality "through doing," I will illustrate manipulative ethical reasoning using a list of invariant behaviors that I call "moral templates," which represent embodied patterns of possible moral behavior, either preexisting or newly created in people's mind-body systems, that enable a kind of moral "doing." I also think it is useful to cognitively compare moral deliberation to diagnosis, a strategy that reveals the logical details of the intrinsic "incompleteness" of knowledge in ethical inferences.

Using a cognitive and epistemological approach to the concept of abduction and model-based reasoning, as I am proposing here, produces an important and valuable side effect: an integrated view that forms a unique framework through which to study the multiple aspects of moral reasoning, including those that are verbal/propositional, model-based, distributed ("moral mediators"), and embodied ("templates of moral doing").

KNOWLEDGE AS DUTY

Chapter 4 is dedicated to explicitly clarifying the motto "knowledge as duty." In our technological world, it has become critically important for us to produce and apply ethical knowledge that keeps pace with the rapid changes around us. We are no less obligated to pursue this knowledge than we are to seek scientific advances; indeed, to neglect the ethical dimension of modern technology is to court disaster. Recent advances have brought about consequences of such magnitude that old policies and ethics can no longer contain them, and we must be willing to approach problems in wholly new ways. Our technology has, for example, turned nature into an object of human responsibility, and if we are to restore and ensure her health, we must employ clever new approaches and rich,

[1] These chapters are largely autonomous and can usefully be read before the other chapters or independently. They are "twins" of a kind and complementary, because they systematically treat similar methodological issues from an epistemological and cognitive perspective (Chapter 7) as well as from a moral perspective (Chapter 6).

updated ethical knowledge. The scope and impact of our current technological abilities have handed human beings the responsibility for, say, "nature" and "the future," which were previously left to God or to fate. Consequently, I declare early in the book – Chapter 3 – my hope for knowledge that maintains and enhances our endowments of intentionality, consciousness, and free-will choices; that strengthens our ability to undertake responsible action; and that preserves our ownership of our own futures. To offer a personal example, while I respect new objects or artifacts that integrate my cognitive activities, I believe it is imperative to explore the moral implications of such devices before embracing their use.

Indeed, basic aspects of human dignity are constantly jeopardized not only by human mistakes and wrongdoing but also by technological products. Constant challenges also come from natural events and tranformations, both ordinary ones, like the birth of one's first child, and extraordinary ones, like epidemics, tsunamis, and hurricanes. I think that preserving and improving the present aspects and characteristics of human beings depends on their own choices about knowledge and morality, and I believe strongly that knowledge is a primary duty that must receive much greater emphasis than ever before and that the knowledge we create must be commensurate with the causal scale of our action. I propose that one way to achieve this and other goals is by accepting "knowledge as duty" and by using disciplines like ethics, epistemology, and cognitive science to rethink and retool research on the philosophy of technology.

What are ethical "reasons"? What is moral progress? What is the role of principles, rules, emotions, and prototypes in ethical reasoning? What is the role of inconsistencies in moral reasoning? Is there a morality "through doing"? These are some of the further issues addressed in the book, along with the practice of casuistry and an analysis of abduction as a form of hypothetical reasoning that helps to clarify processes of "inferring reasons." The book also discusses the problem of free will and examines the role of objects, structures, and technological artifacts as moral carriers and mediators. What all of these topics have in common, though, is that they in some way support the idea that knowledge is our duty. Nearly every thought we have, nearly every action we take, is dictated by the knowledge available to us. In short, if we truly want to effect changes in the world, if we are committed to improving the lives of countless human beings who suffer for a variety of reasons, we must understand that it is only greater knowledge that will allow us to do so.

I started work on this book in 2001, while I was a visiting professor at the Georgia Institute of Technology in Atlanta. In addition to my work here in Italy, I further reshaped the manuscript in 2003 as a Weissman

Distinguished Visiting Professor at the City University of New York, which provided an excellent work environment, and during visits to the Department of Philosophy of Sun Yat-sen University in Guangzhou (Canton), China, where I am currently a visiting professor.

I am grateful to some of my colleagues and collaborators for their helpful suggestions and much more over the last few years, as well as to the two reviewers who read an early draft of the book and whose comments and suggestions helped me greatly as I worked on the final version. I also wish to thank Beth Lindsmith, a freelance editor who enhanced my written English and helped me to frame my ideas in a clearer way.

I discussed many parts of this book with my wife, Anna, to such an extent that I can say we wrote those sections together. This book is dedicated to her, in honor of her idea that ethics are things that help us to become happy.

Pavia, Italy
June 2006

1

Respecting People as Things

Environment

> Now, I say, man and, in general, every rational being exists as an end in himself and not merely as a means to be arbitrarily used by this or that will.
> Immanuel Kant, *Groundwork of the Metaphysics of Morals*

Knowledge, I believe, is fundamental to ethical reasoning, and it must therefore be considered a duty in our morally complex technological world. While I will explore this claim in detail in chapter 4, it is useful to introduce the argument here: if we are to regard knowledge in this new light, we must first understand how knowledge can render an entity moral. To recall our example from the Preface, moral attitudes toward women have greatly evolved over the centuries in Western society, and as our societies have gained greater knowledge, we have ascribed new kinds of value to women. As a result, the cultural default setting is generally that women have an "intrinsic" worth equal to men's.

If acts of cognition can influence moral value, I contend that we can improve the lot of many, many people by altering the way we think about them, and one way to do so is to treat them as things. This notion, of course, flies in the face of Kant's maxim that people should not be regarded as a means to an end – that is, that they should not be seen as "things." As we will see later, however, some things are treated with greater dignity than many people; I argue, consequently, that humankind will benefit if we can ascribe to people many of the values we now associate with such highly regarded nonhuman entities, and to that end, I suggest a new maxim – that of "respecting people as things." In this new ethical orientation, things with great intrinsic value become what I call *moral mediators*: as we interrogate how and why we value such things, we can begin to see how and why people can (and should) be similarly respected. In this way, these things mediate moral ideas, and in so doing they can grant us precious, otherwise unreachable ethical information that will render many of our attitudes toward other human beings obsolete.

RESPECTING THINGS AS PEOPLE

As is commonly known, Kant's categorical imperative states, "*Act only on that maxim through which you can at the same time will that it should become a universal law.*"[1] When dealing with "[t]he formula of the end in itself,"[2] Kant observes that

> man, and in general every rational being *exists* as an end in himself and not merely as a means for arbitrary use by this or that will: he must in all his actions, whether they are directed to himself or to other rational beings, always be viewed *at the same time as an end.* ... Beings whose existence depends, not on our will, but on nature, have none the less, if they are not rational beings, only a relative value as means and are consequently called *things*. Rational beings, on the other hand, are called *persons* because their nature already marks them out as ends in themselves – that is, as something which ought not to be used merely as a means – and consequently imposes to that extent a limit on all arbitrary treatment of them (and is an object of reverence). ... Persons, therefore, are not merely subjective ends whose existence as an object of our actions has a value *for us*; they are *objective ends*, that is, things whose existence is in itself an end, and indeed an end such that in its place we can put no other end to which they should serve *simply* as means.[3]

Kant uses the word "end" in a very formal way, as synonymous with "dignity"; its teleological nature is, after all, not important. Kant is very clear on this point when he writes that "[t]eleology views nature as a kingdom of ends; ethics views a possible kingdom of ends as a kingdom of nature. In the first case the kingdom of ends is a theoretical Idea used to explain what exists. In the second case it is a practical Idea used to bring into existence what does not exist but can be made actual by our conduct – and indeed to bring it into existence in conformity with this Idea."[4] Hence, Kant defines the "kingdom" as a "systematic union of different rational beings under common laws."[5]

These considerations lead us to the following practical imperative: "*Act in such a way that you always treat humanity, whether in your own person or in the person of any other, never simply as a means, but always at the same time as an end.*"[6] In the "kingdom of ends everything has either a price or a dignity. If it has a price, something else can be put in its place as an equivalent; if it is exalted above all price and so admits of no equivalent, then it has a

[1] Kant, 1964, p. 88.
[2] Ibid., pp. 95–98.
[3] Ibid., pp. 95–96.
[4] Ibid., p. 104.
[5] Ibid.
[6] Ibid., p. 96. For Kant, intentions are central to morality. The will is the central object of moral appraisal. Maxims of actions articulate the agent's intentions.

dignity."[7] Things that human beings need have a "market price"; moreover, items that are merely desired rather than needed have an affective "fancy price" [*Affektionspreis*]. But "that which constitutes the sole condition under which anything can be an end in itself has not merely a relative value – that is, a price – but has an intrinsic value – that is, *dignity*."[8]

A simple example involving human moral behavior and responsibility can illuminate the Kantian perspective. Economists say that a decision results in a negative externality when someone other than the decision maker ends up bearing some of the decision's cost. Responsibility is externalized when people avoid taking responsibility for the problems they cause and delegate finding a solution to someone who had no part in creating the trouble. When those who must bear the consequences of a decision are not aware that such a task has been delegated to them, they are treated as means. On the other hand, of course, responsibility is internalized when people accept responsibility for the outcome of their actions.

Kant's wonderful lesson can be inverted: it is possible for things to be treated or respected in ways that one usually reserves for human beings. Many things, or means, previously devoid of value or valuable only in terms of their market price can also acquire moral status or intrinsic value. Conversely, just as things can be assigned new kinds of value, so too can human beings, for there are positive aspects of treating people as things, as we shall see.[9]

A Profound Struggle

Anthropocentric ideas, like those that inform Kant's imperative, have made it difficult for people to acquire moral values usually associated with things and for things to attain the kind of moral worth traditionally reserved for people. As we have said, people should not be treated as if they were a means to an end, but I argue that in some cases we should do just that. My idea for a new maxim – one retooled for the twenty-first century – is, as I have said, to respect people as things in a positive sense. In this scenario, people are respected as "means" in a way that creates a virtuous circle, one in which positive moral aspects enjoyed by things can be used to reshape moral endowments attributed to people, as in the examples I will give in the following chapters.

[7] Ibid., p. 102.

[8] Ibid.

[9] To further clarify my concern about the moral relationships between "people" and "things," the distinction between moral patients and moral agents will be considered in chapter 6 (in the section "Templates of Moral Doing").

Assigning the values of things to human beings seems a bit unnatural, but I believe we can become more comfortable with the concept by analyzing the more familiar practice of ascribing value in the opposite direction – that is, the practice of granting things the value we generally associate with human beings. Attributing moral worth to nonhuman things can be seen as a combination of the Kantian imperative and John Stuart Mill's idea of freedom: "The only freedom which deserves the name, is that of pursuing our own good in our own way, so long as we do not attempt to deprive others of theirs, or impede their efforts to obtain it."[10] If, as Mill teaches, beings (or things, we now add) have a right to something, they are entitled not only to the goal itself but also to the unobstructed pursuit of it. When things also became regarded as entities with interests and rights of their own, the philosophical conceptual space of utilitarianism (animals suffer!) and the idea of environmental ecology were constructed. How did this happen?

One particular type of thing has long been used as a sort of corollary to human beings – animals, whose human-like properties and functions, for example, make possible their use in biomedical research. In this field, studies using *animal models* have induced certain conditions in nonhuman creatures that have allowed scientists to draw conclusions about some human conditions. Researchers achieve such results by exploiting analogies (the fact that rats and humans are alike in various ways, for instance) rather than disanalogies. This theme is very important in philosophy of science because modeling is a widespread scientific practice. Many epistemological problems arise, however, like the challenge of identifying the qualities that make an animal model valid and appropriate.[11] In ethics, however, I contend that the challenge is to look at animals and things not only as scientific models but also as moral models; doing so will help us in our quest to respect people as "means" and to create a virtuous circle that enriches the moral endowments attributed to humans.[12]

ECOLOGY: "THINGS" IN SEARCH OF VALUES

Women, Animals, and Plants

I have said that an entity's value can be recalibrated by knowledge, but how does this process occur? Let us return to the fact that women's intrinsic worth has shifted over the centuries; women are, perhaps, among the most significant "things" to gain new moral rights in Western culture, a change that was not universally welcomed. Indeed, the ideas in this direction

[10] Mill, 1966, p. 18.
[11] Shelley, 2006.
[12] Cf. Magnani, 2007b.

propagated by Mary Wollstonecraft in her 1792 treatise *A Vindication of the Rights of Women* were initially considered absurd.[13] In the last few decades, similar derision has been leveled against animal rights advocates and environmental ethicists, groups that have faced struggles reminiscent of eighteenth-century women's. Just as Wollstonecraft attempted to cast women as beings with great intrinsic value, some intellectuals and activists have sought to reframe how various plants, animals, and ecosystems – even the land itself – are valued so that they are regarded as "ends" and accorded the rights and protection that such status entails.

This way of thinking, of course, could lead to many consequences: if animals are high-status ends rather than means, for example, most experiments on animals would have to be considered wrong. But how should we decide which, if any, organisms are morally suitable subjects for medical research? If it is only a capacity like reason or speech that distinguishes between beings who deserve moral consideration and those who do not, animals would be acceptable subjects, but then so too would infants, the mentally impaired, and the abjectly senile. In this case, classical utilitarianism is the simplest approach to the problem of conducting research: sentience, defined as the capacity to feel pleasure and pain, as a prerequisite for having interests, could be a good alternative criterion. Under this construct, however, many animals could easily acquire moral status: pigs, veal calves, and chicken would therefore have the right to more space in their cages, and experimenting on animals and eating their flesh could be seen as dogmatic speciesism. Moreover, what about plants, soil, water, and air? Are long-lived conscious beings ("Kantian" beings, I would say) intrinsically more valuable than ephemeral or insentient beings? Not necessarily. All living beings have value, but how is that value allocated? From what is it derived?

Various kinds of knowledge and reasoning play roles in assigning new values to animals: (1) anthropocentric arguments, which hold that mistreating animals is related to the possible mistreatment of human beings (a point also stressed by John Locke); (2) utilitarian considerations about sentience and the derived equality of humans and superior animals; (3) ontological notions that, as living creatures, animals and trees have rights in themselves and so are worthy of respect regardless of their effect – positive or negative – on human beings (different ethical gradations exist, of course);[14] and (4) biological awareness of the interconnections among all organisms, objects, and events in the Earth's biosphere.

[13] Singer, 1998.

[14] The individual's right not to be harmed must also be extended to animals, so that killing animals has to be at least "well justified": "Thus, the members of the whaling industry, the cosmetic industry, the farming industry, the network of hunters-exporters-importers must justify doing so [killing whales]. ... Possibly the rights of animals must sometimes

As we see, then, animals' values can be recalibrated by knowledge, and in the following two sections we will see that other entities have been similarly transformed. Exploring these newly valued items will reveal a vast quantity of "things" that we can use as a guide to returning worth to many undervalued people around the world.

Land, Organisms, Species, and Ecosystems

Like individual species of animals, entire ecosystems have, in many cases, also been granted greater value. Biotic communities are as real as human communities: they are dynamic and unstable in the same way, so that morally they can be considered coordinated "wholes." All the individuals – soil, water, plants, and animals – are members of an interdependent community, a "land pyramid," in Aldo Leopold's words.[15] This author, who defines conservation as "a state of harmony between man and land," observes that some biotic communities – marshes, bogs, dunes, and deserts, for example – lack economic value and are generally ascribed less value of other kinds as well. "Unlike higher animals, ecosystems have no experiences; they do not and cannot care. Unlike plants and organisms, ecosystems do not have an organized center and do not have genomes. They do not defend themselves against injury or death. Unlike a species, there is no ongoing telos, no biological identity reinstated over time."[16] Ecosystems do, however, have a "systemic value," and not just because they contribute to human experiences.

The practice of applying ethics to ecological settings and external objects can be seen as a product of evolution itself, that is, either as a biological "limitation on freedom of action in the struggle for existence" or as a "differentiation of social from antisocial conduct."[17] Put another way, we could say that some moral behaviors exist because of natural selection. Our primitive ancestors had no idea that they had genes and, consequently, no interest in their transmission. How, then, did some gene lines survive?

Natural selection rewarded creatures with altruistic feelings, and evolution favored impulses "that originally served to enhance their own genotypic reproduction, but which were deflected to broader social ends in changed circumstances."[18] Thus altruism can be seen as an expression of the selfish gene, but altruistic behavior serves purposes beyond mere

give way to human interests. ... Nevertheless, the onus of justification must be borne by those who cause harm to show they do not violate the rights of the individual involved." (Regan, 1998, p. 539)

[15] Leopold, 1998.

[16] Ibid., p. 138.

[17] Leopold, 1933, p. 634.

[18] Baird Callicott, 1998, p. 154.

self-centered pursuits: is not Leopold's idea of a biotic community related to this kind of biological altruism? Some biologists[19] problematize this relationship between biology and morality and distinguish between biological altruism, which yields collective reproductive benefits, and vernacular altruism, which involves disinterested generosity among human beings. Edward Wilson offers skeptical observations on the biological origins of altruism: it would be a kind of "bounded rationality" of the human brain, he says, that makes simple empathy an efficient rule of thumb. From his perspective, altruism is a logical form of self-promotion or preservation that can be compared to Machiavellian strategies.[20]

Philosophically speaking, Leopold's idea of environmentalism as a product of evolution can itself be seen as related to the Millian "freedom... of pursuing our own good in our own way" without attempting "to deprive others of theirs, or impede their efforts to obtain it" in the sense that we also limit our freedom when we attribute values to external objects, or, in other words, when we practice altruism.

Caring People, Caring Things

Assuming different ethical perspectives is, of course, essential if we are to see difficult issues in new ways and ultimately recalibrate our value systems. Take the field of ecofeminism, for example, whose adherents see women's traditional role of caring for children and "local environments" as part of a mythical matrilineal past, a peaceful agrarian era untainted by the mechanistic modern technology that has now severed the connection between people and nature.[21] As a patriarchal and "rational" worldview has become privileged, the value of both nature and care giving has diminished. For ecofeminists, the argument is simple: women, who, like nature, have been considered "things" for millennia, have an immediate and "organic" relationship with the natural order – that is, with other things – that affords them a more loving, less arrogant perception of the nonhuman world than that held by men. A holistic, spontaneous pluralism is considered a natural component of the female worldview, but attendant characteristics – skill in and a propensity for care giving, for example – have always been considered emotionally centered behavior and therefore outside the realm of traditional ethics. Ecofeminists, however, contend that these very qualities make women ideally qualified to care for both people and their environments and to teach others these skills. The ethics of care is aware of the embodiment of the self and thus of

[19] For example, Sober, 1998.

[20] Wilson, 1998b, p. 486. On the evolutionary origins of moral principles, see also Maienschein and Ruse, 1999, and the recent de Waal, 2006.

[21] On the impact of technology on women, see Bush, 2000.

the importance of things that surround human beings. It is a "local" ethics that functions in private and situated settings and is, as a result, removed from the dominant patriarchal tradition, which values the simplicity of clear-cut abstractions – rules and principles that create an order unburdened by human complexity.[22]

Preserving Things: Technosphere/Biosphere, Human/Nonhuman

Many animals, species, and biotic communities, among other things, are nonhuman entities whose value must be preserved, or in some cases reestablished, and they too can be redefined by learning to think differently about them. Are scientific advances and new knowledge needed to accomplish these goals?

While evolutionary changes are slow and local, human actions can cause sudden massive change; problems often result when human intervention accelerates the normal rate of extinction, hybridization, or speciation. Consequently, it is our duty to anticipate the possible ramifications of our actions. Understanding the scale of potential environmental changes can help us make wiser choices about the ecosystem and its preservation, as is clearly explained by Baird Callicott:[23]*Homo sapiens* must ethically evaluate any changes made to the land to ensure that such projects are conducted on an appropriate scale, thereby minimizing environmental impact. This question of scale is important, for example, when analyzing present-day mass species extinction, a phenomenon that can occur naturally but is certainly hastened by rash human manipulation of the environment. And while long-term atmospheric shifts have occurred for millions of years, the rapid rate of global warming of the last century is abnormal; the environmental change is clearly anthropogenic, as much scientific evidence indicates.

A new moral construct has become necessary because of the tremendous impact that human behavior has on the world: we must now address the issue of the technosphere, that is, the human-made techno-social world in which ecological problems are examined in their social and political contexts. Overpopulation, sustainability, industrial-chemical pollution, social justice, the greenhouse effect, ozone depletion, chemical pesticide use, genetic mutations, and biodiversity destruction are just some of the issues related to the life of the technosphere.[24] In the past – in ancient Greece and Rome, for example – damage to species and to the environment were

[22] Plumwood, 1998. On the role of care giving in ethical knowledge and reasoning, see chapter 6 of this book.

[23] Baird Callicott, 1998.

[24] The religious arguments in favor of protecting biodiversity, mainly related to the story of Noah, are illustrated in Nagle, 1998b, and Kates, 2000.

local and limited, but at present, corporate and megatechnological forces have inextricably linked human beings with wildlife, and the result is nonsustainability at the global level. Population biologists calculate that one or two billion human beings worldwide, living at a basic-needs level of consumption, is the maximum number of people the Earth can support and still maintain ecological sustainability; in 2006, the number is already well past six billion.

In an attempt to deal with such challenges, deep ecology assigns inherent moral value to groups of external things (nonhuman life)[25] that include more than just plants and animals: even things that are usually thought of as nonliving, like rivers (watersheds), landscapes, and ecosystems, attain enhanced value. In this way, Kant's famous maxim undergoes a kind of ethical Copernican revolution: no natural object is considered solely a resource; no "natural" thing can be treated merely as a "means." Natural things do not "belong" to humans, as contended in the traditional anthropocentric view: "Humans only inhabit the lands, using resources to satisfy their vital needs. And if their non-vital needs come into conflict with the vital needs of nonhumans, then humans should defer to the latter."[26] The self-realization of humanity can be reached only if "self" means something very large and comprehensive.

For some people, the free market itself is considered a solution to the environmental crisis. It is well known that politicians and bureaucrats are rewarded for obeying economic pressure groups rather than cooperating with ecological ones. But the market forces that destroyed environments in the past can change attitudes in the future and even address unsolved ecological problems. Such benefits, however, would require that full value – value as end – be given to the property, allowing owners to make decisions from a more fully informed position.

Green-market environmentalism holds that an unregulated market inevitably generates a crisis, and it advocates green taxes to offset the ecological costs incurred. It also asserts that a corporation must assume responsibility for the ecological outcomes brought about by its products, for sustainability requires that prices reflect not only the cost of production but also the cost of repairing any damage to the environment. Many transitions, in fact, must be effected: tax systems must be reformed, linear systems reshaped into cyclical ones, methods of production recalibrated, consumers educated, and human health addressed in ecological terms, to name a few. Donald Fuller[27] contends that the returns of sustainable marketing can be great: the approach can be viewed not as "a pious

25 Naess, 1998.
26 Ibid., p. 202.
27 Fuller, 1999, p. ix.

exercise in corporate altruism" but as a kind of strategy where "customers win (obtain genuine benefits), organizations win (achieve financial and other objectives), and ecosystems win (functioning is preserved and enhanced) at the same time."[28] Of course, the problem of sustainability is exacerbated by the recent trend in business toward commodifying cultural needs without regard for the potential negative effects on human dignity.[29]

Liberal environmentalists, in turn, contend that government regulation must be extended to prevent future environmental damage: the liberal ideal of concern for others can be extended to nonhumans, animals and living objects as well as ecosystems.[30] These environmentalists view any taxation as regressive; if producers are charged for externalities, they will pass on the cost to consumers, and the poor will end up shouldering a disproportionate financial burden. In this view, penalizing the producer is preferable, even if damages are not local and apprehending the criminal is difficult.

Many authors observe that the invisible hand of the market cannot be trusted to prevent ecological crises, and huge, market-driven industries and firms continue practices that are simply unsustainable. It is also maintained that assigning a price to sickness and suffering, not to mention to animal and human life, is a very tricky business indeed; such cost-benefit analysis also discounts the value of future human beings' lives. Environmental imperatives are matters of principle that cannot be bargained away in an economic fashion. Some commentators usefully stress a very interesting paradox of liberalism – that in matters of conservation, one could maintain that neutrality toward others' behavior is necessary to protect their freedom of choice. It is evident, however, that this notion conflicts with the fact that the destruction of natural things limits the freedom of those future human beings who will be deprived of both choices and competing ideas that would have been options had the destruction not occurred.[31]

The problem of continuously destroying natural goods and things, which results from a failure to value them adequately, is illustrated in

[28] Similarly, Kates, 2000, and Flavin, 2000, discuss the "energy revolution" in terms of sustainability.

[29] Nevertheless, is it possible to think of a commodification of intrinsic values like human dignity, in our era of increasing technological and all-encompassing commodification of a large part of sociocultural needs, aspects, and endowments? I maintain that in some sense this could be welcome and good, if related to a respect for egalitarian rights. Intertwining economic relationships with some aspects of human dignity could coincide with a certain degree of social demand and need for them. I will consider this issue in chapter 5, in the section "Commodification of Human Dignity?"

[30] de-Shalit, 1998.

[31] Ibid., p. 398.

Garret James Hardin's famous essay "Tragedy of the Commons," in which he describes the environment as a pasture destined to become a disaster.[32] It is worth noting that John Vogler[33] sees this academic critique as a critical tool in exposing the ineffectuality of the international relations approach to dealing with the commons. Organizations that oversee such activity are incapable of grasping the global socioeconomic phenomenon of environmental change: their "top-down" orientation renders them managerially and bureaucratically blind, and they are guided by attitudes that are "based upon an atomistic science antithetical to ecological holism." They privilege the state in an era of globalization, which impedes behavior that might nurture the commons. Indeed, the globalizing forces of capitalist development, a goal shared by governments seeking economic growth, continually threaten local commons, and a failure to see the connection between local and global commons invites economic collapse.

From a Rawlsian perspective, by contrast, in which hypothetical contracting agents determine the framework of just political institutions and social practices, ecological primary goods can be desired by rational self-interested persons: potable water, safe air, uncontaminated food supplies, future generations' basic needs, and so on.

Of course, in socialist ecology, externalizing the costs generated by global corporations is considered the main cause of many problems – deterioration of the environment, human illness, poor working conditions, social injustice, skewed income distribution, and the mistreatment of minorities. Even assigning a monetary value to environmental "things" is not related to any other kind of value they may have. Moreover, it is certainly possible to give a monetary value to the parts of a whole, but in the case of an ecosystem, ecological value is also a function of the relationships among those parts.[34] It is not a surprise that ecology is in perfect accord with the socialist tradition: the material interchanges between nature and society and within nature had already been stressed by Karl Marx himself when he emphasized the importance of the material conditions of economic production. If we add that material life is also the interchange between nature and human beings, given that consciousness is founded on material life, consciousness needs to be "ecological": "... we need socialism *at least* to make the social relation of production transparent, to end the rule of the market and commodity fetishism, to end the exploitation of human beings by other human beings. We need

[32] Hardin, 1992. On the various problems of global commons – oceans, Antarctica, outer space, the atmosphere – see Ostrom, 1990; Buck, 1998; and Vogler, 2000.

[33] Vogler, 2000, p. 213.

[34] Harvey, 1993.

'ecology' at least to make the social productive forces transparent, to end the degradation and destruction of the Earth."[35]

Also portraying nature as a set of intertwined wholes is the so-called social ecology that comes from the tradition of social geography and ecological regionalism of Reclus, Geddes, Bookchin, and Mumford. In this framework, the self-realization of human beings can be achieved only by respecting nature; consumption, concentrated economic power, authoritarian ideologies, patriarchy, and technocracy – all located outside nature – are evil, and a new imaginative knowledge about nature has to be attained. The political solution to ecological problems is the promotion of communitarianism, cooperation, regionalism, and ecological agriculture. Other so-called bioregionalists, like Snyder and Sale, think that decentralizing both the cooperative economy and the population is necessary in order to solve the ecological crisis.[36]

Current data assessing the human impact on vegetation, animals, soil, water, geomorphology, climate, and the atmosphere are illustrated by Andrew Goudie, who notes a dramatic increase in the complexity, frequency, and magnitude of environmental impacts after the Second World War.[37] This increase is attributed both to steeply rising population levels and to a general increase in per capita consumption.[38]

Ends Justify Means

I have said that knowledge is fundamental in ethical behavior, and I will explain in chapter 4 how knowledge has to become a duty in our current technological world. Various acts of cognition allow things to acquire new cognitive, affective (etc.) values and/or new moral values, and I maintain that we can enhance human value by using the same strategy. "External" things and commodities with desirable values – levels of importance that many people cannot claim – can become vehicles that transfer worth from nonhuman to human entities. When things serve this role, they become *moral mediators*: they mediate moral ideas by revealing parallels between things that are more valued and people who are less valued, thereby granting us precious ethical information and values (other important aspects and cognitive roles of moral mediators will be illustrated in chapter 6).

[35] O'Connor, 1998, p. 415.

[36] The book *Environmental Philosophy* includes a survey of all these so-called political approaches to ecology: see the introduction (Clark, 1998) and part four (Zimmerman et al., 1998).

[37] Goudie, 2000.

[38] Impressive information about ecosystem restoration policies is provided by Egan and Howell, 2001.

Moral mediators, a concept I have drawn from that of the epistemic mediator,[39] are living and nonliving entities and processes – already endowed with intrinsic moral value – that ascribe new value to human beings, nonhuman things, and even to "non-things" like future people and animals. Think for a moment of cities with extensive, technologically advanced library systems in which books are safely housed and carefully maintained. In these same cities, however, are thousands of homeless human beings with neither shelter nor basic health care. Thinking about how we value the contents of our libraries can help us to reexamine how we treat the inhabitants of our cities, and in this way, the simple book can serve as a moral mediator. We are surrounded by human-made and artificial entities, not only concrete objects like screwdrivers and cell phones, but also human organizations, institutions, and social collectives: all of these things have the potential to serve as powerful moral mediators. For this reason, I strongly contend that it is crucial for current ethics to address the relationships not only among human beings, but also between human and nonhuman entities. The concept of the moral mediator also serves as a window on the idea of the distributed character of morality and as such is central to moral reasoning. Later in the book, I will illustrate in greater detail other moral mediators involving the ethical neglect of some environmental external entities and, further on, the amazing case of the so-called endangered species wannabes.

In the previous sections we have seen many cases of how we enhance various things' worth, how we ascribe to them levels of value that we usually reserve for human beings. Now we will examine instances of how value flows in the other direction – that is, from things to people. Consider the following example from Africa in which certain people have been deemed less valuable than some species of animals. Zimbabwe recently instituted a shoot-to-kill policy against poachers of animals that are important to tourism; since the new plan was set in place, dozens of poachers have been killed. Responding criticism of the policy, the head of the antipoaching patrol drew a parallel to events in Europe. He said:

"When a group of Arabs makes an assault on the British crown jewels there is a skirmish and lots are killed – to protect rocks – and nobody minds. Here we're protecting a world heritage, but it happens to be animals and that hangs people up." The moral choice here is complicated by the fact the government has a considerable economic stake in ensuring the survival of the rhinos, since wildlife tourism is big business in Zimbabwe. ... this new "lesser of two evils" problem,

[39] A succinct description of this analogy is also given in the Preface to this book. The epistemological counterpart of moral mediators, which has to do with manipulative abduction, is illustrated in chapter 7.

with its choice between watching a species become extinct or using violence to prevent it, is likely to confront society in different forms again and again in years to come.[40]

Because of their emotional impact, I think that practices such as ecotage and monkey wrenching[41] can help us to understand "things" as "moral mediators," and that in turn can lead us to precious, otherwise unreachable ethical information and help us to return value to human beings in a variety of situations. For example, radical environmentalism contends that ecotage belongs to the tradition of defending minorities because unprotected plants and animals being destroyed "are" minorities. Moreover, if the considered self is ecologically extended, as many contend, and becomes part of a larger ecological self comprising the whole biological community, the defense of "things" becomes "self-defense," and, consequently, breaking the law is justified. Acts of ecotage compel our culture to face the fact that it currently considers a bulldozer (and the related profit it represents) of higher value than a living and intact ecosystem composed of a diverse community of plants and animals. "In the future, as more and more species become extinct and forests are recognized for their role in maintaining a livable biosphere, this value system may be judged equal to such historical extravagancies as the burning of women suspected of witchcraft or the internment of loyal Asian-Americans during World War II."[42]

When people of the future, who will undoubtedly live in an era of ecological scarcity, look back on the early twenty-first century, more than a few of our actions, laws, and policies will be seen as horrifying lapses of both intelligence and ethics. When the pernicious consequences of deforestation, the greenhouse effect, and ozone depletion result in a world with one season – an endless hot summer – the ethical neglect of external things as moral mediators will be clear.

Under present law, a timber company can purchase the trees in an old growth forest, cut them down (often at taxpayers' expense), and leave the forest biome so disrupted that the animal and plant communities it previously supported perish or migrate. Erosion may make local streams unsuitable for the spawning of salmon and therefore affect the ocean food chain hundreds of miles away. Pursued on a large scale, as is certainly happening today, the fragmentation of forests will increase global

[40] Manes, 1998, p. 462.

[41] This is a term from Ed Abbey's book *The Monkey Wrench Gang* (1975). The novel concerns the use of sabotage to protest environmentally damaging activities in the American Southwest, and was so influential that the term "monkey-wrenching" has come to mean, besides sabotage and damage to machines, any violence, sabotage, activism, law making, or law breaking to preserve to wilderness, wild space, and ecosystems.

[42] Ibid., p. 461.

warming and inevitably lead to a higher rate of extinction, as the findings of island biogeography demonstrate. All this is totally legal. By contrast, those who try to preserve the forest by spiking the trees are guilty of vandalism under the present law.[43]

INTRINSIC AND INSTRUMENTAL VALUES AS MORAL MEDIATORS

We have said that knowledge actively renders entities moral, as it has enhanced the moral status of women and animals, inanimate objects, and technological structures at various times in history. At issue in such transformations is the concept of *intrinsic value*, or the worth an entity has for its own sake, and this idea is central to my approach: analyzing how people have ascribed intrinsic value to various entities is more and more urgent given the important role of moral mediators in the lives of modern people. Ronald Dworkin, citing Hume's subjectivism, reminds us that some philosophers deny the very possibility that anything has an intrinsic value.[44] Indeed, viewed from David Hume's perspective, objects and events can be valuable only when they serve someone's or something's interests, that is, if they have "instrumental" value. The following are two fundamental questions posed by Dworkin: "How can it be important that a life continues unless that life is important for or to someone? How can a life's continuing be, as I am suggesting, simply important in and of itself?" This worth results from our treatment of some objects and events as valuable in themselves; we say that they have an independent, *intrinsic* value. We have endowed things like books, flags, paintings, and ecological systems with intrinsic value; so too have we ascribed such worth to more amorphous entities like human cultures.

> We treat the value of someone's life as instrumental when we measure it in terms of how much his being alive serves the interests of others: of how much what he produces makes other people's lives better, for example. When we say that Mozart's or Pasteur's life had great value because the music and medicine they created served the interests of others, we are treating their lives as instrumentally valuable. We treat a person's life as subjectively valuable when we measure its value to him, that is, in terms of how much *he* wants to be alive or how much being alive is good for him. So if we say that life has lost its value to someone who is miserable or in great pain, we are treating that life in a subjective way.[45]

We consider all human beings to have intrinsic value whether or not they also have instrumental or subjective value: human life is important in

[43] Ibid., p. 459.
[44] Dworkin, 1993, p. 69.
[45] Ibid., pp. 72–73.

and of itself. It is in the light of this conviction that abortion acquires a problematic moral status. Dworkin introduces a further distinction: things that already enjoy intrinsic (or "sacred," as he says) value also have *incremental* value. Knowledge, for example, typically has incremental value – the more of it we have, the better off we are. But human life is intrinsically valuable just because it exists.

Association and designation are two ways to make something intrinsically valuable: in ancient Egypt, for instance, cats were revered because they were associated with certain goddesses; today, flags are respected because of their link to the life of certain nations. Other objects become "sacred," as Dworkin says, because of their history. "It is not what a painting symbolizes or it is associated with but how it came to be that makes it valuable. We protect even a painting we do not much like, just as we try to preserve cultures we do not especially admire, because they embody processes of human creation we consider important and admirable."[46] The same scenario applies to some animal species and ecological systems. We do not aim to protect them for our own pleasure, for people who consider endangered species important may never encounter any of the animals they want to protect. Instead, we seek to preserve them just because we think it would be a shame if human acts and decisions were to destroy them. As with human life, such animals and ecosystems do not hold incremental value: "... few people believe the world would be worse if there had always been fewer species, and few would think it important to engineer new bird species if that were possible. What we believe important is not that there be any particular number of species but that a species that now exists not be extinguished by us."[47] Life forms that exist independently of people are more valuable than the ones we have genetically engineered: we do not consider plants created by scientists to be intrinsically valuable in the way that naturally produced plants are.

Both species of life and works of art are regarded as expressions of complicated acts of creativity: evolution or God created species, and human beings works of art; and these things must survive and prosper so that future generations can be possible and can enjoy these natural and cultural resources.

When intrinsic values have *degrees*, our judgments about them are *selective*: the painting of a great artist has greater value than one of a minor artist, but a rare, exotic bird may be more important to us than an endangered but deadly snake whose extinction might engender less regret. I agree with Dworkin when he says that we treat human-made art as inviolable, but we withhold such esteem from automobiles and

[46] Ibid., pp. 74–75.
[47] Ibid.

commercial advertising, even if people also create these. Similarly, we do not treat everything produced by a long natural process – coal, diamonds, petroleum deposits – as inviolable either, and many people willingly cut down trees to clear space for a house or eat complex mammals like rabbits and cows for food: "... few people care when even a benign species of insect comes to an end, and even for those who believe that viruses are animals, the eradication of the AIDS virus would be an occasion for celebration untinged by even a trace of regret."[48]

Degrees of value and selectivity result from the way different bodies of knowledge – scientific, philosophical, religious, popular – have historically depicted the intelligibility of things and living creatures. Also, in the case of human beings, we consider the intrinsic value of life to be related to age: value would slope upward "from birth to some point in late childhood or early adolescence, then follow a flat line until at least very early middle age, and then slope down again toward extreme old age."[49]

Degrees of value and selectivity are also historically changeable, subject to intellectual and scientific revolutions and transformations. Christian theologies almost always envision a universe with humankind at the center, making it difficult to attribute intrinsic value to animals (notwithstanding Saint Francis's exception). Surely, modern biological knowledge, at least since Darwin, has established a new intellectual environment in which to consider animals, plants, and other natural things. It was the seventeenth-century biological theory of homunculus that visualized the fetus as a miniature adult, enabling us to regard it as a person endowed with all the moral value of an adult. This viewpoint persists despite various scientific and theological convictions (Saint Augustine's, for example) that consider an embryo to have nonhuman status, at least in the first weeks.

Can things so endowed with intrinsic value serve as moral mediators that elucidate precious new ethical information and values that would otherwise be unknowable?

THE DISPOSSESSED PERSON, THE MEANS-PERSON, AND THE THING-PERSON

The Dispossessed Person

Because of the intrinsic value of human life, the premature deaths of human beings are a frustration of their own and others' expectations. Dworkin uses the word *frustration* to describe a premature death because of the investments that the deceased and others have made in that

[48] Ibid., p. 80.
[49] Ibid., p. 87.

truncated life – individual and shared ambitions, plans, expectations, love, interests, and personal involvement.[50]

Frustration, then, cannot be associated with a mere absence of life; it is predicated on a life that exists and then ceases to exist: it is easy to see how this perspective is important in the case of the debate on abortion (as well as in discussions of suicide, assisted suicide, and euthanasia). If we think frustration is greater when a person dies before he or she has made significant personal investments in life, then smaller degrees of frustration can be hypothesized in the case of early-term abortions, when embryonic growth has just started and the natural "investment" is still small. In this example, it is clear that we attribute more importance to the creative side – the *human* and *social* side, so speak – of the intrinsic value attributed to human life, as in the case of the liberal perspective on abortion. In this light, it could be accepted that the future life of a deformed fetus is more frustrating than the immediate death caused by abortion. On the contrary, conservatives think that only the *natural* side of the intrinsic value attributed to life is important (and not just because the fetus has rights and interests), so that abortion is always and a priori wrong.[51]

Frustration happens not only because of death but also because of many other kinds of failure, as Dworkin says, like handicaps, poverty, misconceived projects, irredeemable mistakes, bad luck, and lack of training. Such incidents can be seen as random events that are not caused by the actions of other people, but I would add that frustration is, of course, also the result of harm and disrespect that must start somewhere. Human beings are disturbingly adept at inventing ways to dispossess others of their right to pursue a full and flourishing life. Moreover, it is not accidental that in collectives, there are always ongoing conflict negotiations and continual revisions to the "list" of things and events considered immoral and/or frustrating in this sense.

The Means-Person

A means-person is a person who, not surprisingly, is treated as a means to an end, or who is, as Kant says, "for arbitrary use by this or that will." He also declares that "[m]an ... must in all his actions, whether they are directed towards himself or towards other rational beings, always be viewed *at the same time as an end*."[52] Here I will offer four Kantian examples, the first three of which relate to the importance of humanity as an end in itself and the fourth to the need for promoting it.[53] Promoting

[50] Ibid.

[51] Cf. Feinberg, 1992; Dworkin, 1993; Rachels, 1999.

[52] Kant, 1964, p. 95.

[53] Ibid., pp. 97–98.

humanity requires active involvement, for abstaining from acting and/or deciding can be a form of treating people as if they were mere means. As Mill writes, "A person may cause evil to others not only by his action but by his inaction."[54]

(1) Contemplation of suicide is incompatible with the idea of humanity "as an end in itself." If a person contemplates suicide, he or she aims to escape a painful situation by using a person (him- or herself) merely as a means to maintain a tolerable state of affairs until the end of his or her life. (2) A man who makes a false promise to others intends to "make use of another man *merely as a means* to an end he does not share"; the deceived man cannot possibly agree with the first man's way "of behaving to him and so cannot himself share the end of the action." (3) "[I]n regard to contingent (meritorious) duty to oneself, it is not enough that an action should refrain from conflicting with humanity in our own person as an end in itself: it must also *harmonize with this end*. Now there are human capacities for greater perfection which form part of nature's purpose for humanity on our person. To neglect these can admittedly be compatible with the *maintenance* of humanity as an end in itself, but not with the *promotion* of this end." (4) "As regards meritorious duty to others, the natural end which all men seek is their own happiness. Now humanity could no doubt subsist if everybody contributes nothing to the happiness of others but at the same time refrained from deliberately impairing their happiness. This is, however, merely to agree negatively and not positively with *humanity as an end in itself* unless every one endeavours also, so far as in him lies, to further the ends of others."[55]

As already indicated, Kant believes that in the "kingdom of ends everything has either a *price* or a *dignity*. If it has a price, something else can be put in its place as an equivalent; if it is exalted above all price and so admits of no equivalent, then it has a dignity. ... Things," as related to general human inclinations and needs, have a "market price. ... What is relative to universal human inclinations and needs has a market price; what, even without presupposing a need, accords with a certain taste ... has a *fancy price* [*Affektionspreis*]," but "that which constitutes the sole condition under which anything can be an end in itself has not merely a relative value – that is, a price – but has an intrinsic value – that is, *dignity*." Skill and diligence in work have a market price; wit, lively imagination, and humor a fancy price; "... but fidelity to promises and kindness based on principle (not on instinct) have an intrinsic worth. In default of these, nature and art alike contain nothing to put in their

54 Mill, 1966, p. 17. See also chapter 5, this volume, the section "Immorality and Lack of Morality."

55 Kant, 1964, p. 102.

place.... Therefore morality, and humanity so far as it is capable of morality, is the only thing which has dignity."[56]

In the following chapter, I will explore the practice of cloning and discuss how it relates to the problem of treating people as means. Here, however, it is sufficient to note that the market economy is inherently inclined to treat people as if they were means, to exploit them without regard for their own aspirations in a way that benefits others or enhances relatively abstract "external things," like economic growth, the destiny of the nation, the reputation of an institution, and so on.

The Thing-Person

I contend, of course, that voluntarily impeding others' ability to develop and use their intellectual or physical qualities is a frustration of their expectations. Let us assume that the intellectual qualities of a human being qualify him or her as worthy of moral consideration. Thinking in terms of cognitive capacities, a human being could be likened to an artificial "thing" that is the bearer of information, knowledge, cultural traditions, and so on, and thought of in the same way that we might regard objects worth moral consideration: a book, a PC, or a work of art, for example.

Let us return to the idea of the library and consider the life of a typical library book: depending on its age and value (not only instrumental and economic), librarians record its circulation, monitor its condition, repair it when needed, and replace it when necessary; books in wealthy countries are generally guaranteed such treatment. But the same care is not extended to many people who are carriers of the same knowledge one might find in the book just described or in other external objects like databases. Unfortunately, the cognitive content and skill of human beings are not always given the same rights and moral value as a book or a database. There are no precise moral (and/or legal) rules that enjoin us to tend to the cognitive skills of human beings or the information they carry in the same way that we care for external objects and configurations endowed with cognitive worth.

Like books in a well-managed library, many organizations and institutions – nations, universities, and businesses, for example – are also endowed with intrinsic values that are absolutely unimaginable for people. I argue that we must ascertain which of those values can also be attributed to human beings so that they too can enjoy the types of moral worth that are now associated only with external things. In these circumstances, we face a paradoxical situation in which people are not "sufficiently" or appropriately treated as means, or things, in an anti-Kantian sense. I think

[56] All quotations are from Kant, 1964, p. 102.

human beings must be respected as we now respect many things; people deserve, at the very least, the same level of care we now lavish on library books. In this way, "things" serve as moral mediators (cf. chapter 6, the section "Moral Mediators") that show us how to assign new values to human beings. By examining the intrinsic moral worth we have collectively delegated to things, as in the case just described, we can learn how to enhance the moral value of fellow human beings.

Many things that surround us are human-made, artificial – not only concrete objects like a hammer or a PC, but also human organizations, institutions, and societies. Other entities that fall into this category include economic markets, laws, corporations, states, and school structures. As I illustrated earlier, we have also projected many instrumental and intrinsic values onto those things (or parts of them) – flags, for example, courtroom trial rituals, ecological systems. These external objects and configurations have also established a kind of automatism that conditions us and distributes roles, duties, and moral engagements.

As human beings become moral clients, so too do nonhuman things (and, I would add, so-called non-things like people and animals of the future), and consequently, current ethics must address not only the relationships among people, but also those between human and nonhuman entities. While human beings exhibit many of the same qualities as objects, the two groups obviously do not share every characteristic. For example, some objects' economic value engenders a kind of intrinsic value that largely overcomes the value of a life; similarly, attributing intrinsic moral value to the market rights of objects can, in many cases, overshadow human economic and moral value. But people too are potentially endowed with economic value (quantifiable, as insurance companies know) and cognitive worth. Hence, they are moral carriers as well as emotional repositories, knowledge carriers, and so on.[57] Curiously, these endowments we share with "things" are often disregarded simply because, paradoxically, we are "just" humans – not "external things." How can we counter such discrimination?

RESPECTING PEOPLE AS THINGS

Endangered Species Wannabes

Delegating moral features to external things, or "nonhumans," in an ecological framework sometimes stirs up a miasma of human dissatisfaction. Every day we see external things – a building, a cultural tradition, or a view, for instance – being endowed with economic and/or inherent moral value, but there is relatively little interest, in, say,

[57] Cf. the last section of this chapter, "Humans as Knowledge Carriers."

denouncing the exploitation of some people by others in various work settings. Industries externalize their costs of preserving the environment onto people – customers and others – who come to feel burdened, frustrated, and disrespected. I think a solution to this paradox might lie in the *moral mediator*, which can show us how and why to value people as much as we do external things that already enjoy considerable instrumental and inherent worth.

Moral mediators are often used to add value to a variety of entities; they are frequently prominent factors, for instance, in the interesting process of identifying and managing endangered species.[58] Numerous species have acquired intrinsic moral value, which has led to their being legally classified as endangered in many countries. According to a recent report[59] of the United States Fish and Wildlife Service, there are 1,424 endangered species, both plant and animal, that are entitled to some impressive legal protection. Membership criteria for this protected group, however, have been interpreted as discriminatory and limited, and some groups have demanded that certain people, places, and other things also be considered "endangered." Nagle[60] describes an ontologically diverse and unbelievably long list of subjects aspiring to the title: New England fishermen, the California taxpayer, middle-class citizens, ranchers, farmers, loggers, infantrymen, corporate middle managers, manufacturing workers, private doctors, park rangers, shrimpers, peanuts, sugar, Atlantic fisheries, American-made typewriters, the maritime industry, amusement park rides, public television, old songs and stories of the Acadian community in Maine, young African-American males, free white human beings in New York, the Jordanian state, women in India, Tibetans, Democrats, family relationships, morality, "housewives and nothing more," African-American judges on U.S. Courts of Appeals, cultural traditions, unborn children, and so on. The list is hilarious, and it is even more surprising to learn that some of these subjects have actually been granted endangered species status in judicial opinions. But in each case, advocates have used (or attempted to use) acknowledged endangered entities as moral mediators to add value to entities related to their chosen causes.

Many people have complained about disappearing wildlife receiving more moral and legal protection than disappearing cultural traditions. A recent federal statute, the Visual Artists Rights Act of 1990, appropriates the language of ecological preservation when it establishes "rights of attribution, integrity, and the prevention of destruction of art of

58 Nagle, 1998a.
59 <http:/www.fws.gov/r9endspp/boxscore.html>.
60 Nagle, 1998a.

recognized stature for the creators of certain paintings, drawings, prints, sculptures, or photographs."[61]

Such efforts to draw parallels between endangered species and other entities are attempts to validate their (mainly, but not exclusively, moral) importance and to obtain some sort of legal protection for them. Of course, the threat of extinction is the only legitimate qualification for endangered species status, and it is hard to see the disappearance of typewriters as a profound loss to humanity. Not all things are worth saving, of course, but we have learned something new by examining how people seek to redefine as "endangered" something or someone they see as threatened.

The importance of this analogy is that some people consider themselves endangered because they do not feel as if they are treated as well as things (means). And there are many cases in which people should be considered uniquely valuable in the way we now value endangered species. Going beyond the humorous instances just listed, human characteristics like cognitive attitudes and the ability to intelligently manipulate the world qualify people as repositories of knowledge with the capacity to reason and work, and they must therefore be considered unique resources to be preserved and enhanced. Few agencies now undertake this protective role: while many organizations exist to protect things that have informational value, it seems that no one defends people as valuable carriers of knowledge. I think such a lack of support explains our failure to acknowledge human dignity in our technological world.

Finally, of course, even if we wanted to save everything, to do so is impossible, and priorities have to be established even among endangered species recognized by the ESA (Endangered Species Act, U.S., 1973).[62] Nagle describes the problematic of the decisions adopted by the so-called God Squad that originated the act: it is impossible to protect all species, and limited resources make choosing among threatened life forms inevitable. Utilitarian criteria (the usefulness of a species as a source of medicine, food, esthetic pleasure, or tourism opportunities, for example, or as a necessary component of an ecosystem as a whole) are very controversial and subject to constant cultural, social, and political negotiation. Some religious criteria would seem to solve the problem by attributing an intrinsic worth to every species, but doing so creates an impractical and unwieldy value system.

Dehumanizing Humans

Reintegrating humanity into the natural world,[63] as the environmental philosophers do, means in some sense dehumanizing and renaturalizing

[61] Ibid., p. 249.
[62] Nagle, 1998b.
[63] Kirkman, 2002.

humans. A tacit assumption could be that human beings, who are capable of choices and actions – if they are free, that is, and, consequently, anti-natural – can decide to abandon a bygone state of harmony with nature.[64] This anthropocentrism, when mixed with ignorance and arrogance, can affect environmental equilibria. We have clearly seen that environmentalism attributes to "things" a certain intrinsic moral value (and legal rights, too) previously held only by humans. At the same time, human beings have sought values and claimed characteristics that have traditionally been associated only with things, qualities like naturality and a "primitive" belonging to the whole of nature: human subjectivity and natural objectivity are both encompassed by the supposed "subjectivity" of an organic wholeness;[65] in this way, the standard ontologies are modified.

We already know that freedom belongs to the realm of Kantian transcendence,[66] and so it is abstracted from the materiality of human biological determination. On the other hand, some biologists, who tend to see most human problems in terms of their science, maintain that freedom is a kind of biologically adaptive illusion shaped by evolution.[67] From the environmentalist point of view, human freedom is just one aspect of the spontaneity of nature, and, as we have already observed, ethics are a product of an evolution that favors human cooperation.[68]

Reifying Humans and Humanizing Things

Once intrinsic moral value is attributed to external "things," as in the case of endangered species, for example, those "things" limit human choices and create new problems. We have already described this idea as an extension of the Millian "freedom ... of pursuing our own good in our own way" without attempting "to deprive others of theirs, or impede their efforts to obtain it." By attributing values to "external objects" as well as to other "people," we increase the number of entities we must respect when making choices, thereby decreasing our freedom.

The environmentalist views of the above ontologies raise to a new level the usual ethical, legal, and political problems concerning regulations among human beings. "Things" become humanized, and human beings, in turn, are reified as it becomes clear how they affect and depend on

64 On human beings as choosers and responsible agents, see chapter 4, the section "Identity and Privacy in the Cyberage," as well as chapter 5.
65 Kirkman, 2002, p. 33.
66 Kant, 1964.
67 Wilson, 1998a, and chapter 3 of this book.
68 Leopold, 1998.

these objects. Because more comprehensive scientific and ethical knowledge is now at our disposal, we – human beings – will be able to make better decisions.[69]

Human and nonhuman beings are deeply integrated, so much so that it becomes profoundly reductive to see problems only in terms of the selfish economic needs of human global capitalism. While they hold various conceptions of nature and of their relationship with nonhumans, all humans are involved in the natural world. This idea springs not just from the "metaphysical" datum that declares that as living bodies we share the destiny of nature and of other things (necessity) and that as free spirits we share the realm of transcendence (freedom); it arises simply from the fact that, in ways we are not yet able to appreciate, we are cognitively and morally integrated with other beings, so that any changes we inflict on the world are also inflicted on ourselves.

HUMAN AND NONHUMAN COLLECTIVES, HUMAN AND NONHUMAN AGENTS

Humans and nonhumans are inextricably intertwined: "You are different with a gun in your hand; the gun is different with you holding it."[70] We are in a sense "folded into" nonhumans, so that we delegate action to external things (objects, tools, artifacts) that in turn share our human existence with us. The idea of the "collective" expresses an exchange of human and nonhuman properties akin to what I have just described in the case of things in search of intrinsic value: "what the modernist science warriors see as a horror to be avoided at all costs – the mixing up of objectivity and subjectivity – is for us, on the contrary, the hallmark of a civilized life."[71]

Many such examples are mentioned by Bruno Latour: he cites instances in which knowledge about nonhumans is used to reconfigure people and, conversely, to project the properties of humankind onto nonhuman entities. When considered from the ethical perspective, the first case reflects our problem of respecting people as things, while the second reflects ideas illustrated earlier in the chapter about moral representations of nonhuman: "The new hybrid remains a nonhuman, but not only has it lost its material and objective character, it has acquired properties of citizenship."[72] Of course, the nonmoral case of endowing nonhumans entities with speech, intelligence, and other human properties – things from classical media to computational tools, from paintings

[69] Cf. chapters 4 and 6 of this book.
[70] Latour, 1999, p. 179.
[71] Ibid., p. 200.
[72] Ibid., p. 202.

to artificial intelligence, from a simple tool like a hammer to a sophisticated machine – is related to this movement. So too are agriculture and the domestication of nonhuman animals, which involves their socialization and reeducation.

In turn, external things (the electrical, transportation, and telecommunication industries, for example) have constructed new social frameworks, and factories, machines, and institutions have created previously unknown roles for people by establishing new forms of management constraints and human negotiation: "It was from techniques, that is, the ability to nest several subprograms, that we learned what it means to subsist and expand, to accept a role and discharge a function."[73] Tools, which have always served as human prostheses, become integrated into our bodies as we use them in a kind of anthropological transformation of both the individual and the collective. This mixture of human and nonhuman is also expressed in human bodies that are increasingly shaped and integrated by "sociotechnical negotiations and artifacts."

The cyclical process of transferring qualities between humans and nonhumans is, of course, an inextricable part of our evolution, and consequently it requires ongoing negotiations and a continual redrawing of the lines between the two kinds of entities.

Latour's notions of the dehumanizing effect of technology are based on the so-called actor network theory.[74] The actor network theory basically maintains that we should think of science, technology, and society as a field of human and nonhuman (material) agency. Human and nonhuman agents are associated with one another in networks, and they evolve together within these networks. Because the two aspects are equally important, neither can be reduced to the other: "An actor network is simultaneously an actor whose activity is networking heterogeneous elements and a network that is able to redefine and transform what it is made of. ... The actor network is reducible neither to an actor alone nor to a network."[75]

A different but related perspective – one that, like Latour's, avoids anthropomorphic prioritization of human agency and addresses the dissolution of boundaries between things and people – is offered by Andrew Pickering in his writing on post-humanism. He describes externalities (representations, artifacts, tools, etc.) as kinds of nonhuman agencies[76] that interact with a decentered human agency in a dialectic of

[73] Ibid., p. 209.

[74] This theory has been proposed by Michel Callon, Latour himself, and John Law (cf. Callon, 1994, 1997; Latour, 1987, 1988; Callon and Latour, 1992; and Law, 1993).

[75] Callon, 1997, p. 93.

[76] As a form of what Pickering calls "disciplinary agency," nonhuman agency also includes conceptual tools and representations – scientific theories and models, say, or mathematical formalisms: "Scientific culture, then, appears as itself a wild kind of machine built from

"resistance" and "accommodation" called *the mangle of practice*.[77] Resistance is a failure to capture material agency in an intended form, while accommodation amounts to a reconfiguration of the apparatus that might find a way through its resistance. When human and nonhuman agencies are brought together, as has often occurred in mathematics, the natural sciences, and technology, it is impossible to predict the results.

Composite human/nonhuman agents – "cyborgs" and "sociocyborgs" – are protean beings, operating as they do in the always-shifting realm of science and technology. From this perspective, scientific practice is "organized around the making (and breaking) of associations or alignments between multiple cultural elements," and "fact production, in particular, depends on making associations between the heterogeneous realm of machinic performance and representation, in a process that entails the emergent mangling and interactive stabilization of both."[78]

HUMANS AS KNOWLEDGE CARRIERS

Because technology is so rapidly changing, it may seem impossible to foresee its future impact on human beings, but classical philosophy offers a way to anticipate the problems that await us. The issue is interesting and worth discussing at the end of this chapter, which deals with the interplay between beings and things. In the *Grundrisse*, Karl Marx discusses the possibility of capitalistic collectivities' evolving in a positive way, an evolution that can result from the increasing role of science and machines in production, and he foresees the development of what he called the "general intellect." He describes a future in which the "creation of real wealth" will depend "less on labour time and on the amount of labour employed than on the power of the agencies set during labour time" and will be contingent instead on "the general state of science and on the progress of technology, or the application of science to production." Then, he continues, labor will no longer appear "so much to be included within the production process," and the human being will "relate more as watchman and regulator to the production process itself." Marx's

radical heterogeneous parts, a supercyborg, harnessing material and disciplinary agency in material and human performances, some of which lead out into the world of representation, of fact and theories" (Pickering, 1995, p. 145).

77 Ibid., p. 17 and pp. 22–23.

78 Ibid., p. 29 and p. 159. Pickering's theory also provides an interesting critique of the aforementioned "actor network theory," which he critiques as being too inclined to exclusive semiotic considerations of human and nonhuman agencies. For this reason, Pickering thinks the theory misses both the "temporal" dimension of emerging material nonhuman agency as well as important asymmetries between the human and material realms, like the presence of intentions and sociality.

statements seem to predict today's decrease in labor that has resulted from the computer revolution.[79]

In this transformation, the human being performs "his understanding of nature and his mastery over it by virtue of the presence of a social body – it is, in a word, the development of the social individual which appears to be the great foundation-stone of production and of wealth." We will reach a point where we will have reduced "the necessary labour of society to a minimum; which then corresponds to the artistic, scientific, etc. development of the individual in the time set free, and with the means created, for all of them." Marx is perfectly aware of the fact that artifacts – or, in his words, "machines, locomotives, railways, electric telegraphs, self-acting mules" – are "*organs of the human brain, created by the human hand*; the power of knowledge, objectified." He concludes with the following: "The development of fixed capital indicates to what degree general social knowledge has become a *direct force of production*, and to what degree, hence, the conditions of the process of social life itself have come under the control of general intellect and been transformed in accordance with it."[80]

The general intellect is a collective social intelligence created by accumulated knowledge, techniques, technologies, and know-how. In the process described by Marx, the hybridization of human and machine is a pivotal event in the recent history of economic production; it is so important that it can be situated at the center of the constitution of the human being. In his vision, people are enriched with intellectual and cooperative power in unprecedented ways, and this anthropological transformation occurs through new kinds of work made possible by rapidly changing technologies.

Human beings are no less important than the nonhuman artifacts to which they, as hybrids, are so closely related. Because we already respect nonhuman artifactual repositories of knowledge – libraries, medical machines like PETs and MRIs, computers, databases, and so on – it should be easy to learn to respect human ones; we need only expand our idea of "knowledge carrier" to include people in that category. The widespread hybridization of our era makes it necessary – but also easy – "to respect people as things."[81]

How ethical knowledge is attributed to external things needs to be clearly understood; so too does the process of transferring values back to human

[79] See also Rifkin, 2000.

[80] All quotations are from "*Contradiction between the foundation of bourgeois production* (value as measure) *and its development. Machines etc.*," Notebook VII, 1858 (Marx, 1973, pp. 704–705).

[81] Cf. chapter 4, this volume, the section "New Knowledge Communities and Humans as Knowledge Carriers."

beings. It appears that neither of these two goals can be reached through traditional moral philosophy and that if we are to succeed, we must expand our current approaches by studying the cognitive mechanisms that effect such shifts. The ways in which we have managed to morally reclassify other entities – women, species of animals, and ecosystems, for example, as we have seen – can serve as a guide to replenishing the value of many people. We must also strive to understand our relationship to tools, artifacts, and other objects, external things to which we have delegated various actions, for the human and the nonhuman have become so inextricably intertwined that we share both a common existence and a common set of values. Further knowledge can be gained by comparing highly valued external repositories and carriers of knowledge to people who are themselves caches of important skills and information but have been denied the worth we have heaped upon computers, databases, and the like. The values we hold are always the result of the knowledge we have, and the knowledge we have always arises from the cognition we engage in. Cognition, then, is a powerful tool indeed.

2

Treating People as Means

Cloning

> It still remains unrecognized, that to bring the child into existence without a fair prospect of being able, not only to provide food for its body, but instruction and training for its mind, is a moral crime, both against the unfortunate offspring and against society; and that if the parent does not fulfil this obligation, the State ought to see it fulfilled, at the charge, as far as possible, of the parent.
>
> John Stuart Mill, *On Liberty*

The concept of "respecting people as things" provides an ethical framework allowing us to analyze many aspects of the modern human condition, like the medicalization of life and the effects of biotechnologies. I contended in the previous chapter that the modern undermining of Immanuel Kant's distinction between instrumental value (based on ends and outcomes) and intrinsic value (ends in themselves) results from the blurring of traditional distinctions between humans and things (machines, for example) and between natural things and artifacts. I am convinced that this shift in thought in some sense contradicts the Kantian idea that we should not treat people as means, a notion often cited by those who are suspicious of biotechnology.

Indeed, when one condemns human cloning, one usually appeals to Kant's moral principle that a person should not be treated simply as a means to other people's ends; an example might be the biotechnological cloning of a human being solely for use as a bone marrow donor. This kind of argumentation underlies many declarations against cloning and other kinds of biotechnology, and all of these objections are linked to the central problem of human dignity. To use people for cloning would be to trample on their autonomy; to create unneeded embryos during in vitro fertilization (IVF) would be an instrumentalization that manipulates human beings.

These ideas, however, do not hold up under all moral frameworks. Using Steinbock's writing on Dworkin's theory of intrinsic value, I derive the proposition that entities like clones and embryos do not, in fact, have a full moral status, but that they do deserve a "profound respect" and "serious moral consideration" and are therefore endowed with a certain degree of intrinsic value, even if they are not considered complete human beings. Because they are not fully human, Kant's exhortation does not apply to them, but my proposed precept to "treat people as things" does: this particular moral status should not disqualify unneeded frozen embryos from use in "relevant" medical research (genetics, cancer, etc.) or in cloning; destroying them in the process of pursuing these goals is therefore acceptable. Other considerations also support the fact that the Kantian maxim is irrelevant to human cloning, which, in many respects, resembles other forms of assisted reproduction. I contend that his maxim can be applied only in circumstantial cases.

Even if, as Kantians fear, such technology makes it impossible to avoid treating people as means, and cloning technology does indeed have the potential to produce "monsters," all is not lost: it is certainly possible to build new ethical knowledge to manage these new entities and situations. Moreover, those people, as amalgamations of the human and the artificial, will clearly seem more "thing-like" than the type of person we have encountered throughout history. Does this not heighten the urgency of learning to "respect people as things"?

XEROX COPIES, RIGHTS, HUMAN AND NONHUMAN DIGNITY

When one rejects cloning,[1] as we have said, one usually appeals to Kant's moral principle that people should not be treated simply and solely as a means to an end. It is also contended that cloning violates human dignity, even in the context of infertility treatment, where, unlike the cloning scenario just described, there is no clear instrumental role assigned to the resulting clone.[2] Rarely in human collectivities do people regard one another as "ends in their own right," as human beings whose happiness is important in itself, independent of any positive impact on others. As

[1] A cloned embryo is made by taking an enucleated oocyte (an egg with the nucleus removed) and inserting into it the nucleus of a somatic cell, one extracted either from the same organism or from another organism of the same species. The oocyte then nurses the transplanted nucleus until it is fully integrated into its new location, and if all goes well, it will develop into a full-grown being. Cloning can be considered a concrete manipulation that involves moral conflicts and problems; in this sense, it is also a kind of action-based moral problem. I will illustrate the issues related to this kind of morality in the section "Being Moral through Doing: Taking Care" in chapter 6.

[2] Putnam, 1999.

I shall explore more fully in chapter 4, this lack of respect also fuels the problem of privacy.

Does cloning violate the "dignity of human beings" and the "security of human genetic material," as the director of the World Health Organization maintained in May 1997? In March of the same year, the EU Parliament proclaimed that cloning is also "a serious violation of fundamental human rights and is contrary to the principle of equality of human beings as it permits a eugenic and racist selection of the human race," and, moreover, that "it offends human dignity and it requires experimentation on humans." In Europe, especially, there is a notable emphasis on human dignity when dealing with the problem of cloning.

How could the security of genetic material be compromised? John Harris asks, "is it less secure when inserted with precision by scientists, or when spread around with the characteristic negligence of the average of human males? ... Appeals to human dignity are, of course, universally attractive; they are the political equivalent of motherhood and apple pie. Like motherhood and apple pie they are comprehensively vague."[3] Just to offer a reassuring example, we know that the dignity of a natural monozygotic twin is not threatened by the existence of his brother.[4]

Even more radically, Richard Lewontin observes that it is hugely hypocritical to worry about the Kantian maxim as long as the relationships of capitalistic production exist: "The very word 'employment' and 'employee' are descriptions of an objectified relationship in which human beings are 'things' to be valued according to externally imposed standards."[5]

Clones as Means/Things

As the fruit of hybridizing humans and artificial techniques, clones will obviously seem to have more thing-like qualities, to use an expression from the previous chapter, than other people who, presumably, came into being the old-fashioned way. Hence, what is the moral status of these hybrid beings? Are they persons in all the same respects as we are? I do not think so. Are they thing-like people? Again, I do not think so. In this and the following sections, I will attempt to explicate their moral status: by the end of the chapter, we will see that our motto "respecting people as things" is very useful in producing a better understanding of their moral configuration.

[3] Harris, 1999, pp. 65–66.

[4] Some people would object that there is a fundamental difference: identical twins are contemporary, whereas clones are not (on the problem of cloning and of the future of human reproduction, see Harris, 1998; Harris and Holm, 1998; and Wachbroit, 2000).

[5] Lewontin, 1997, quoted in Putnam, 1999, p. 8.

The Kantian objection to treating others as mere stepping stones to one's goals implies that using human clones would violate their autonomy, their ability to live according to their own desires and values (on the problem of owning our own destinies, see chapter 3). We can treat people as means, but we need not treat them *exclusively* as means – we can, at the same time, also regard them as ends. Both cloning and in vitro fertilization,[6] in which spare embryos are created, are usually thought of as kinds of instrumentalization that manipulate human beings. The assumption here is that embryos are human beings and, as such, enjoy all related rights, including the right to life and the Kantian right not to be used as "mere means."[7] This construct also includes the right not to be subjected to experiments that carry the risk of death but offer no compensating benefit. In these cases, sentience is considered the ability to experience pain and pleasure and constitutes the discriminating condition for having any interests at all. Working from this definition, nonsentient beings, whether mere things (like radios, armchairs, or artificial apparatus) or living organisms without nervous systems (like vegetables), do not have interests of their own. Bonnie Steinbock employs Dworkin's theory of intrinsic value, which was illustrated in the previous chapter, when she makes the following remarks:

> This is not to say that they cannot be cared for, fed, protected, repaired, and healed, or killed, injured, made sick, and destroyed. It is rather to say that it does not matter to the non-sentient beings whether we care or not. We can preserve their existence, and even promote their welfare for their own sake, out of concern for their feelings, for they do not have any. The view that links moral status to interests and restricts interests to sentient beings I call *the interest view*. ... The claim is not that we should be concerned to protect and preserve only sentient beings, but rather that only sentient beings can have an interest or a stake in their own existence.[8]

Of course, nonsentient beings have value, but the destruction of an important work of art cannot be understood in terms of how it directly affects the work of art.

Very early embryos cannot experience pain, and Steinbock contends that unless there is a clear awareness of some kind, a being does not

[6] Usually considered the "port of entry" for human embryo research.

[7] An interesting opposing position that supports clones' special status is advocated by the U.S. President's Council on Bioethics. In a recent *New England Journal of Medicine* article, Paul McHugh contends that because clones do not result from unions between eggs and sperm, they should be considered what he calls "clonotes" rather than "embryos." This framework removes any moral concerns about disrupting a naturally produced embryo, and, consequently, he contends that it renders cloning research morally acceptable (McHugh, 2004).

[8] Steinbock, 2001, p. 23.

possess a life to lose. Killing does not necessarily mean taking a life; spermicide kills millions of sperm, for example, but it is absurd to say that they are losing their lives. From this perspective, it seems that it is conscious existence rather than biological life that matters most, and that if we kill a nonsentient embryo we deprive it of nothing. In this sense, an embryo has a biological life but not (yet) a biographical life.[9] The "interest view" proposes that "it is prima facie wrong to deprive beings of their biographical lives but not wrong to end merely biological lives, of course when there are good reasons for doing so."[10]

We know that some ethicists, however, see rights as inherent rather than contingent upon sentience, and they consider it irrelevant that embryos lack consciousness. From this perspective, abortion is wrong – as wrong as killing an adult human being. But even the construct in which rights derive from sentience does not legitimize killing a temporarily unconscious person.

However, it is necessary to distinguish between, say, a person in a coma, who has had beliefs, interests, and desires in the past, and a fetus that has never experienced such things. Maybe we can say that its future is its interest, but we cannot say which future belongs to the fetus; to ascertain this, it is necessary to apply a theory of identity that usually requires a degree of psychological continuity not present in the early fetus: "Although brain waves are present at about 6 weeks after conception, more development of the cerebral cortex is necessary for there to be awareness of painful stimuli, arguably the most basic type of awareness. The neural pathways are not sufficiently developed to transmit pain messages to the fetal brain until 22 to 24 weeks of gestation."[11]

Do conjoined gametes (and the subsequent embryo) have a future like ours (FLO)? Unless an individual gamete is involved in fertilization, it has no future; left alone, it will die, and Steinbock maintains this is "also true of the embryo that cannot develop all by itself." Killing an embryo cannot be wrong, then, because it is impossible to deprive it of its future. On the other hand, this lack of a personal future does not mean that abortion is justified: the fact that a being (the embryo) lacks moral status and rights does not automatically warrant its destruction. A highly simplistic, less fraught analogy might be that one cannot reasonably kill a barking dog just because it is annoying. The admissibility of abortion arises from two factors: the moral status of the unborn as well as the pregnant woman's right to bodily self-determination.

Instead of asking if the fetus is in fact a person, another way to interrogate the abortion issue is to ask whether killing an embryo and killing

[9] This distinction has been introduced by the moral philosopher Rachels (1985).

[10] Steinbock, 2001, p. 34. See also Steinbock, 1992.

[11] Steinbock, 2001, p. 26.

an adult are wrong for the same reasons. I will not address this question in detail because it is less pertinent to the themes I am exploring, but it is worth a brief look. Killing a "normal" adult is wrong because it deprives him or her of a FLO, but in killing a fetus we are depriving it of a future life, too.[12] Moreover, as I have illustrated in the previous chapter, this approach also stresses that intrinsic value lies in the "natural" side of existence – simply being alive – and is, as a result, more important than any "human" or "social" aspects of personhood. If intrinsic value derives from the state of being alive rather than from any fetal rights and interests, then abortion, is always and a priori wrong.

By contrast, following Steinbock's interest view, the embryo does not have a moral status but does, it is said, deserve "profound respect" and "serious moral consideration," as maintained in the United States by the Ethics Advisory Board (EAB) in 1979 and by the Human Embryo Research Panel of the National Institute of Health (NIH) in 1994.

Some external objects and things are endowed with intrinsic value because of the associations they suggest – art, wilderness, cultures, species, collectivities, or even, as Joel Feinberg remarks, a human corpse. From this perspective it would be wrong, for example, to hack up Grandfather's body after his natural death and dispose of his remains in the trash can:

> That would be wrong not because it violates Grandfather's rights; he is dead and no longer has the same sort of rights as the rest of us, and we can make it part of the example that he was not offended while alive at the thought of such posthumous treatment and indeed even consented to it in advance. Somehow acts of this kind if not forbidden would strike at our respect for living human persons (without which organized society would be impossible).[13]

To continue this line of thinking: if we do not respect dead bodies, it is likely we do not respect the living. Analogically, disregard for embryos suggests a similar antipathy, for, like Grandfather's body in Feinberg's example, an embryo becomes a "symbol of human life." The moral status of a symbol, however, should neither impede the use of unneeded frozen embryos in medical fields like cancer research[14] nor preclude their use in cloning; the destruction of surplus human embryos can be permitted, even justified, by this moral attitude. It would be better to donate the extras, but the rights of the gametes' donors can override the attempt to

[12] See, for example, Marquis, 2003.

[13] Feinberg, 1992, p. 53.

[14] Nevertheless, a symbol's moral status should impede an embryo's use in, say, some cosmetics research, which could be deemed insufficiently important to justify the use of embryos.

preserve the embryos, which would not exist without the consent of the donors.

Cloning is made possible by somatic cell nuclear transfer (SCNT), which has other applications and benefits unrelated to the replication of entire organisms: it yields information about cell growth, division, and specialization that we cannot obtain in any other way, and it offers advances in disease treatment, organ creation, and skin transplants for burn victims.

> The objections to cloning human beings rest largely on the possibility of harm to them. Although this remains controversial because of an ongoing debate about whether a person can be harmed by a technique on which his or her very existence depends, most people would regard risk of serious defects in offspring as an obvious reason to ban the technique. Yet this concern about cloning refers to harm to future persons and has nothing to do with respect for embryos.[15]

Engendering a child through cloning solely in order to produce a son and heir certainly seems to fall under the Kantian criticism, because it treats people as means. John Harris, however, thinks this is not the case, because when we "create" people, "so long as existence is in the created individual's own best interests," the motives for which the individual was generated are "either morally irrelevant or subordinate to other moral considerations."[16]

As already remarked, it is not a surprise that a great part of human relationships is based on commercial – primarily instrumental – relations. In this case, all we need is a list of criteria for the civilized treatment of children in general, cloned or not. There is no reason to believe that future cloned children will not be loved for themselves in a civilized way: "monsters," or "Xerox copies," if we prefer, will still need ethics because they will still be "people" (see the section later in this chapter "Monsters Will Still Need Ethics").

Hence, the Kantian maxim does not apply to human cloning, just as it is irrelevant to other forms of assisted reproduction. Nor does it seem applicable to circumstantial cases of parents who use cloning to create a child "solely" in order to have an heir or a little brother for Jane or to become eligible for public housing.[17] Indeed, parents' choices in designing their children should be completely unrestricted. This freedom has always existed naturally through the choice of the procreational partner, and it has become even broader through widespread reproductive

[15] Steinbock, 2001, p. 33.
[16] Harris, 1999.
[17] Ibid., p. 76.

techniques such as egg donation, sperm donation, surrogacy, and abortion. Harris concludes:

> If they [parents] use cloning techniques for reproduction they may be less surprised by their children's physical appearance, but they will, for sure, be surprised by their children's dispositions, desires, traits, and so on. ... they may be [will be] less surprised by one thing: they should be less unpleasantly surprised by genetic diseases and defects, for they will not only know much about the nucleus donor, but will have had opportunity to carry out genetic tests before creating the clone.[18]

Genetic Variability, Abuse, Uniqueness

Many other issues arise from the prospect of cloning, and all must be analyzed in light of the central problem of human dignity. We have seen that using people for cloning would involve disrespecting their autonomy; the same happens in the case of in vitro fertilization, where spare embryos are created. These actions are usually thought of as a kind of instrumentalization that manipulates human beings. In the following passages, I will provide an analysis of such puzzling issues, of aporias, and of possible consequences of cloning and other biotechnologies: the problems of genetic variability, abuse, uniqueness, and legal aspects will be considered, as well as the role of sex in the future.

Will human cloning reduce human genetic variability with catastrophic results? UNESCO refers to the preservation of the human genome as protecting "the common heritage of humanity." Normal reproduction constantly varies the genome, so it seems that cloning would be the best way to preserve it. Moreover, a technique's potential for abuse does not constitute an argument against the technique. The following argument by Harris is simple but compelling: "to ban cloning on the ground that it might be used for racist purposes is tantamount to saying that sexual intercourse should be prohibited because it permits the possibility of rape."[19]

Most of the dangers attributed to human cloning are those that threaten the physical and psychological welfare of children who have the same DNA as another individual. Humans cloned from an adult cell, as was Dolly, the sheep born in 1997, might have a higher risk of cancer or premature aging. Moreover, as Gina Kolata illustrates in her history of cloning, Dolly was the result of 277 attempts to fuse an adult nucleus with an egg.[20] Of these 277, 27 embryos developed normally in the first week,

[18] Ibid.
[19] Ibid., p. 80.
[20] Kolata, 1998.

but only one developed to term. It seems that 50 percent of the fetuses were lost in the last two-thirds of development, as compared with a 6 percent loss during that period in natural procreation. In addition, 20 percent of the lambs born alive died shortly after birth. Finally, some of the fetuses were abnormal. In summary, it seems that current cloning techniques still present great risk.

Beyond issues of cloning's feasibility, other problems lurk: Will cloning result in a new category of people? Would it lead to a new relationship between the created and the creator? Will a lack of genetic uniqueness violate the principle of human equality? We know that a kind of natural human cloning occurs when, for unexplained reasons, more than one fetus develops from a single fertilized egg, resulting in identical twins. We are also perfectly aware of the importance of nurture in the development of genetically "identical" siblings: genes are a major force in how we turn out, but they are not the only influence that shapes us, which is why the indeterminability of individuals is unaffected by cloning. Indeed, genetic identity is not a central component of personal identity. Research on monozygotic twins clearly shows that while people who share the same genetic constitution may be very similar in many respects, they nevertheless differ sufficiently to leave no doubt about their being two different people. There is little evidence indeed that a lack of genetic individuality will negatively affect human dignity.

Nasal Reasoning, Olfactory Moral Philosophy, and Legal Aspects

David Hume attributed a certain importance to the role of feelings in judging what is morally permissible, an approach that would explain why some critics characterize cloning as disgusting. Harris says this kind of "nasal reasoning" accompanied by its big brother, "olfactory philosophy," recalls the kind of thinking that automatically triggers disgust in people – in racist whites, for example, when confronted by black people. It is a "wisdom of repugnance," as Harris so eloquently puts it.[21]

As we will see in chapter 6 ("in the section Model-Based Moral Reasoning"), we cannot always trust our emotions when dealing with issues of morality. If we fail to permit human cloning on the basis of emotional feelings, we might violate principles of dignity related to the right to procreate without having any compelling reason to do so.

Following John Robertson's point of view, some aspects of human-cloning research must be legally protected.[22] Tradition emphasizes the importance of both individual and family autonomy, notions supported by legal principles and widely held ideas of fundamental rights. This

[21] Kass, 1999.
[22] Robertson, 1998, 1999a, 1999b.

tradition completely legitimizes couples' procreative liberty and, therefore, their interest in raising biologically and genetically related offspring. According to this construct, reproductive and therapeutic cloning that can help form and maintain families would have to be allowed and protected once techniques are found to be sufficiently safe and efficacious. Then cloned embryos could be used to treat infertility, to avoid the need for donor gametes, to serve as a source of organs or tissues, or even to replace a dead child.

Less acceptable, however, is cloning by heterosexual parents in order to select the offspring's genome when (less predictable) sexual reproduction is still possible, or by homosexual parents hoping for a genetically related child. In these situations, cloning collides with the mentality of our present collectivities because of the exaggerated parental control over a child's genome. A final emotional sticking point relates to how we should regard an individual with only one genetic parent, a condition that could be considered unnatural and therefore disturbing. But the number of one's genetic parents – whether one or two – is just one particular component of our view of what constitutes our humanity, with no greater importance (and arguably less) than other factors like rationality, consciousness, and language.[23]

If cloning were to become an acceptable and common form of technology, numerous social transformations would be possible.[24] For example, the practice of cloning appears to be the quintessence of a queer model: sex will be without reproduction; transsexuals and people with AIDS will be able to reproduce; sperm – and the men who produce them – will become unnecessary; having queer children will perpetuate queer culture; the apotheosis of some forms of lesbian feminism will become more likely; same-sex marriage will be allowed. Other possible cloning-related issues are anticipated by other authors: unequal endowments could lead to financial and market transformations; the sex ratio could shift, increasing the power of women; new tax-related problems could occur; a need for surrogate and artificial wombs will arise; and the birth rate of the infertile could eclipse that of the fertile.[25] Berry points clearly to potential legal problems and consequences related to in vitro fertilization and genetic enhancement, such as damage caused by incorrect (or incorrectly conducted) procedures, emotional distress arising from tension between a kind of inauthentic false self and a "real self"; a social imperative encouraging genetic "upgrading" of offspring; and the risk of eugenics.[26]

[23] On the various tendencies of the current research in biotechnology, see Baldi, 2001.
[24] Eskridge and Stein, 1998.
[25] Posner and Posner, 1998.
[26] Berry, 1998, 1999.

Reproductive Technologies

It is absolutely certain that new reproductive practices have the potential to profoundly alter reproduction and parenthood as well as the processes of women's liberation.[27]

> Reproductive technologies have thrown the 'begat-begat' generational model of the genetic lottery into disarray. The hierarchical species model with human beings at the top followed by animals and plants has also been shaken as we discover new homologies between all species, use transgenic plants and animals as vectors for pharmaceutical products and vaccines, and bioengineer pigs to produce human organs. These radical transformations have launched us into a *bio-society*, or, according to its critics, a *biotechnocracy*.[28]

There are many conflicting attitudes about reproductive interventions,[29] and the following are just a few examples. Some believe that these technologies are defensible as long as they respect the well-being of both the offspring and the adults involved and honor individual women's agency in creating, sustaining, and bringing forth life. Men, however, could become superfluous, because they would either have to buy an egg and rent a womb or find a generous woman. The fact that males do not gestate babies could render them obsolete. The liberal claim that people have the "right" to clone themselves is meretricious nonsense – indeed, civilized peoples have long set limits on procreative behavior by banning incest, polygamy, and other forms of "reproductive freedom." While human organ farms are horrifying, we must remember that unmodified nature is not always benign: disease and death are natural, and we seek comfort and health through artificial cures.[30]

Going beyond reproductive technologies, it is also important to mention electronics-oriented biotechnologies that combine advances in prosthetics with computer science. These modern treatments, which restore and rehabilitate human bodies, continue a long tradition established when crutches and peg legs were invented. The effects of these technologies are significant: there are already three million people in the world living with artificial implants.

In the near future, it is likely that micro-electrodes will be implanted in the visual cortex (not only of the blind) so that the brain is stimulated to see scenes captured by a miniature camera – it will be a kind of prosthetic cortical implant. Some people and researchers contend that such implantations threaten the integrity of the human body and degrade

[27] Wajcman, 2000.
[28] Gardner, 1999, p. 29.
[29] Beauchamp, 1999; Beauchamp and Childress, 1994.
[30] Krauthammer, 1999.

human dignity; we must remember, however, that many people already accept invasions of their organic bodies by mechanical apparatus – pacemakers and artificial hip joints, to name two – for curative purposes. As the number of enhanced humans increases through biotechnology and bioelectronics, the current idea of normal will seem subnormal, so additional areas of life will become medicalized.

Other shifts will involve issues of identity and the mind-body dichotomy. For example, the boundary between the physical self and the perceptory/intellectual self will change as new enhancements alter how we perceive and interact. And some ethicists fear that electronic prostheses could be equipped with tracking devices, or something like that, making it easy for future governments to control and monitor citizens. In labor environments, employers and insurers who have genetic information about individuals would be able to discriminate against them based on genetic factors. Of course, expanding reproductive technologies will lead to the commodification of human body parts and even bodily functions – surrogate motherhood, sexual services, genetic enhancement, cloning, and so on.[31]

As new electronic prostheses are developed, new ethical issues will arise. Problems concerning risk, appropriateness, societal impact, cost, and equity must be considered in a multidisciplinary framework of computer science, biophysics, medicine, law, philosophy, public policy, and international economics.[32] Tom Beauchamp and James Childress outline four groups of general principles for future bioethics: (1) respect for autonomy, the decision-making capacities of autonomous persons; (2) nonmaleficence, the idea of avoiding harm to others; (3) beneficence, guidelines for the provision of benefits and the weighing of benefits against risks and costs; and, (4) justice, a set of principles for fairly distributing benefits, risks, and costs.[33]

It is not my intent to discuss these principles in detail, but it seems to me that they generally correspond to judicious ethical ideas that can guide us as we deal with new technology. I simply think that we need not fear advances in biotechnological research, that we should, instead, adopt an enlightened attitude, as I will try to illustrate in the following sections. In order to keep pace with scientific progress, we must continually produce new policies and new ethical knowledge that will expand and fine-tune the four groups of principles and allow us to manage new cases and situations.

[31] In chapter 5, I will treat in detail the problem of commodifying human features and actions in the section "Globalization, Commidification, and Wars."

[32] Maguire and McGee, 2001.

[33] Beauchamp and Childress, 1994. See also Beauchamp, 1999.

The products of cloning and of other reproductive technologies, then, are hybrid blends of human beings and artificial techniques and in a sense are the complex result of the hybridization that began when people first devised simple machines. As medical and other technology grows ever more advanced, human beings will become increasingly integrated with nonhuman artifacts and technical procedures, as would be the case with the imminent "products" of reproductive technology. Consequently, the dictum to "respect people as things" becomes increasingly urgent, and Kant's maxim seems more and more dogmatic and abstract. "Respecting people as things" helps us to consider the new products, insofar as they are "hybrid people," as entities in search of new ethical understanding and new knowledge about their problems and needs.

In the following chapters, as I consider ethical issues like freedom, privacy, responsibility, and equality, I will return to Beauchamp's and Childress's four groups of principles. Also to be explored are the main threats of what Beauchamp and Childress call autonomy and the ideas of free will, consciousness, and owning our own destinies.

MONSTERS WILL STILL NEED ETHICS

It is unlikely, I believe, that anyone will be able to halt the march of technology and stop the creation of "monsters," to use some people's term for future humans created through cloning and other biotechnologies. It is inevitable that human sexual nature will be more hybridized – not only because of the new reproductive technologies we have discussed, but also because of the biotechnological and artifactual enhancement of sexual desire and performance. These facts do not necessarily mean, however, that the future is bleak, for we can build new ethical knowledge that will replace obsolete ethical maxims that are too abstract and general to be helpful and that will help us adapt to technological change. If we create and apply new frames of reference, the sexual and biological "monsters" with new post-human bodies can be seamlessly integrated into society.

Sex in the Future

Sex in the Future, a book by Robin Baker about the "reproductive revolution and how it will change us," examines the possible long-term effects of all reproductive technologies.[34] Baker addresses the decline of the nuclear family, for example, and the trend toward encouraging single mothers and single fathers to consider various forms of cohabitation in order to save money, parental energy, and time. In these one-parent

[34] Baker, 2000.

families, men will have no choice but to help support their genetic offspring, and for the first time in human evolution, conception will become as costly for a man as it is for a woman. Infertility will cease to exist for both women and men, thanks to IVF, surrogacy, artificial insemination, surrogate testes and ovaries, and cloning itself. It can be hypothesized that those families will be more stable than the ones with children who are genetically linked to both parents because there will be fewer differences and less conflict.[35] Darwin too, from a biological perspective, pointed out the paradoxical character of sexual reproduction: a female who spontaneously switched to clonal reproduction would immediately be twice as successful as her sexual rivals.[36]

Out of a world population of more than six billion people, forty-eight million are already clones: we call them identical twins, but genetically speaking, they are as alike as clones. Those who condemn genetic replication must remember that such natural clones have not suffered a loss of dignity from sharing the genes of another person. As Baker notes, there can be little doubt that cloning will become one of the many reproductive options available to future generations. Because the word "cloning" has sinister mad-scientist connotations, accepting this inevitable technology may be easier if we use a neutral, less emotionally fraught term, like "artificial twinning," for example:

> The knee-jerk opposition to cloning in the 1990s was very reminiscent of the opposition to artificial insemination in the 1930s, and IVF technology in the 1970s, and in fact is not so very different from the opposition to wet-nursing in the seventeenth century. ... Cloning offers the ultimate solution to the problem of human infertility, and as such, it cannot fail but to have its day.[37]

I agree with Baker that infertility is a profoundly urgent biological problem and that cloning will be the ultimate solution.

Baker also envisions some provocative scenarios regarding contraception of the future, which will be markedly different from today's options. Contraception has been practiced in various ways since prehistoric times: attempts to avoid pregnancy are evident in records of ancient societies, and there is evidence of such behavior even among prehistoric peoples – indeed, members of some hunter-gatherer populations typically had only three children in their lifetimes. In the future, Baker predicts that a kind of contraceptive cafeteria will freely offer a wide range of options and will facilitate family planning for both women and men. Families, arguably, are likely to have only two children if

[35] See also Rao, 1999.

[36] Dawkins, 1998, p. 57.

[37] Baker, 2000, pp. 158 and 160.

current trends continue – in Western countries there has been a noticeable decline in family size and an increase in the number of women who delay starting their families. It seems that a shop similar to the "cafeteria" just mentioned increased the use of contraceptives in Bangladesh from 7 to 40 percent in a decade.[38] The storage of gametes in any kind of cryopreservation – the BlockBank (BB) system is one example – will enable young men to store a supply of their semen, have a vasectomy, and avoid unwanted pregnancy until the time is right for them to have children, at which point they would retrieve their sperm and use some form of reproductive technology. Giving men and women more control over when and if they have babies is likely to reduce the number of both unwanted pregnancies and abortions.

In addition to birth control convenience stores, there will also be "reproduction restaurants" and a "gamete marketing board" that will allow people to plan their babies in a very "unemotional" way. Parents will be able to choose their children's features, order eggs and sperm, and arrange surrogate mothers. This practice will deepen the disconnect between having sex and reproducing and return human beings to a view not unlike that of our early ancestors: to them, the two events seemed so unrelated that many "had more trouble making a connection between the two than separating them."[39] We must also add that natural selection itself has already caused a separation between sex and reproduction. It seems that lions, for example, have sex many, many times in order to generate each lion, and some birds engage in similar excesses.

This new reproductive climate might mean that eugenics will start to play a role in day-to-day decisions, but Baker contends such fears are unwarranted. In Western industrial societies, at least, preferences for boys and girls seem to be more or less balanced. After all, selecting a mate can be considered a form of eugenics; we seek in a mate those qualities we want in our offspring. Potential eugenic choices of the future (avoiding offspring with a congenital disease, for example) are in principle – and biologically speaking – no more or less eugenic than choices based on the "implicit knowledge" that comes from physical attraction. Moreover, biotechnology will be able to do what matchmaking was never able to do: provide the means to rid the human population of genetic diseases, at least those caused by a single gene.[40]

Fidelity, infidelity, and promiscuity will acquire new meanings. Infidelity, for example, will no longer be equated with adultery (because sexuality will be completely separated from reproduction): traditionally,

[38] Ibid., p. 193.
[39] Ibid., p. 200.
[40] Ibid., p. 206.

uncertain paternity has heightened men's anxiety about their partners' infidelity:

> Natural selection has shaped people to react badly to the image of their loved one having sex with another person. And it shaped the reaction to be largely subconscious and hormonal, rather than a cerebral calculation of costs and benefits. Nevertheless, even the single parents of the future, temporarily in love with their current sexual or live-in partner, might well still suffer from their evolutionary legacy if that person has intercourse with somebody else.[41]

Another trigger for jealousy has been the possibility of contracting diseases: eradicating them (and/or increasing successful treatment) will end that kind of emotion.

Scientists will develop techniques of genetic alteration by inserting or deleting genes: people say that genetic manipulation could reduce the richness of the human genome; genes for minorities – homosexuals, for example – could be eliminated from sperm and eggs and subsequently disappear. But if we consider homosexuality largely genetic, as Baker does (and indeed, the orientation is also present in many animals), it is unlikely to be wiped out by the new reproductive technologies: probably less than 1 percent of men are exclusively homosexual; 5 percent of men in industrial countries are bisexual; and almost all lesbians (80 percent) have sex with men. "Homosexuals will wish to use future reproduction technology as much as anybody, and in the future they will have a variety of options, one of which will be to reproduce with each other."[42]

To many people, a future involving cloning, surrogate mothers, and IVF may seem unnatural, but these are only among the most recent of many unnatural human achievements: all technology, in fact, relates to something unnatural. On this point, the "biologist" concludes:

> Natural selection shaped identical twins whereas science and technology shaped clones, so we oppose clones. But natural selection shaped humans to walk whereas science and technology allows them to fly, yet we welcome planes. Natural selection shaped humans who succumbed to heart disease, but science and technology via organ transplant and artificial parts often allow them to survive, yet we welcome transplants.[43]

While traditional medicine has extended the life of (and increased reproduction by) many people with various diseases and genetic deficiencies, the result is a genome that is increasingly susceptible to infectious and other diseases in later life: the result is a kind of *dysgenics*: "The

[41] Ibid., p. 230.
[42] Ibid., p. 260.
[43] Ibid., p. 308.

ladle that best mixes the gene pool is global migration, which is ironic given that it was the very thing feared by the racist eugeneticists of the past. In the United States they believed that immigration and interbreeding were increasing genetic defects in the American population."[44]

The human gene pool could be threatened by the Human Genome Project, but genetic variety is unlikely to be endangered by more radical forms of gene therapy as long as individuals are granted freedom of choice and are not forced or directed by repressive organizations.[45] The issue is controversial: other authors maintain that "the mixing of genes that results from sexual reproduction may enhance survival, even under the environmentally gentler conditions brought about by modern medicine."[46]

To conclude, knee-jerk opposition to new technology – the responses based on the "nasal reasoning" and "olfactory philosophy" described earlier – could be strong, but I think this "wisdom of repugnance" is often highly superficial and/or hypocritical. As Baker says on the last page of *Sex in the Future*, "even if people do go back to nature from time to time, they will rarely go far from their mobile phone."

The Right to Reproduce

People in the past did not have a "real" right to reproduce. It is interesting to note that the analysis of cloning (and of noncoital reproduction in general) has also generated a debate on the "right to reproduce" and the responsibilities involved with reproduction the old-fashioned way. Coital reproduction sometimes imposes heavy costs (diseases, unwanted offspring) on members of society other than the parents, and these costs outweigh the personal meaning experienced by those who reproduce.[47] John Stuart Mill considered bringing children into being without the prospect of adequate physical and psychological support as nothing short of a moral crime:

> It still remains unrecognized, that to bring the child into existence without a fair prospect of being able, not only to provide food for its body, but instruction and training for its mind, is a moral crime, both against the unfortunate offspring and against society; and that if the parent does not fulfill this obligation, the State ought to see it fulfilled, at the charge, as far as possible, of the parent.[48]

[44] Ibid., p. 213.

[45] On free will and free choice, see the following chapter.

[46] Posner and Posner, 1998, p. 254.

[47] Robertson, 1999b.

[48] Mill, 1966, p. 129.

Throughout most of history, those who could reproduce did, and those that could not, did not. However, thanks to biotechnology almost everyone will be able to reproduce – if they have access to and the means to afford that technology. Someday financial limitations will become more important than biological infertility in dictating who can and cannot reproduce, a situation that will be exacerbated – or ameliorated – by government policies.

Finally, the right to reproduce is strongly related to biological constraints. Some critics maintain that would-be parents should not be obsessed with having genetically related offspring; these parents, they say, should instead be satisfied with having a child for the child's sake, since identity is shaped not only by genes but also by social and emotional factors. As Baker contends, however, this is a strange and futile claim to make, equivalent to wishing that people had evolved with two heads rather than one. Evolutionary theory holds that all parents of all species prefer their offspring to be genetically related to them, and empirical data confirm that all animals do in fact behave according to this hypothesis: It matters to them whether the child is 'their' child or somebody else's. The very fact that people demand wherever possible to have their own genetic child shows the depth of feeling involved. ... So who has the right to tell anybody that genetic parenthood should not matter?[49]

Post-Human Bodies

Surely millennia of natural selection have rewarded both human and nonhuman animals that prefer genetic offspring. It is also clear that ongoing cultural changes lead us to challenge – then transform – the conventional norms of corporeal and sexual relations between and within genders. Like attitudes, bodies themselves transform and mutate into new post-human bodies, where even an electronically mediated sexual contact will be possible: a related branch of virtual reality has been eloquently dubbed "teledildonics."[50] The first condition of this corporeal transformation is recognizing that human nature is in no way separate from nature as a whole, that there are not fixed and necessary boundaries separating humans, animals, and things.

Ethics will be especially important for artificially contrived post-human monsters,[51] beings who could be considered the result of the human tendency to treat others as means. Their specific problems and relationships will require new forms of ethical knowledge to help them

49 Baker, 2000, p. 315.
50 Rheingold, 1991.
51 See Russo and Cove, 2000.

navigate through life. But perhaps they will need less ethical guidance. Is it not possible that, when looking at the past, they will consider *us* to be the monsters for our willingness to harm people, animals, and the environment?

We must deal with the eventual hybridization of humans and artificial techniques, which will prove to be a transformation so profound that it will be situated at the center of human constitution. Because it appears that human beings will become increasingly integrated with nonhuman artifacts, unnatural objects, and procedures, as the future of reproductive technologies seems to indicate, the idea of "respecting people as things" becomes especially urgent. The changes we wreak on external things affect us immediately because we are integrated with them (IVF, cloning, genetic enhancement, etc.); we must therefore individuate new moral knowledge and identify the possible consequences – both long- and short-term – of our actions. Critics warn that cloning technologies will make it impossible for us to avoid treating people as means, but if we shed the obsolete notions that limit our thinking, we can acquire new knowledge and reasoning skills that will imbue these hybrid beings of the future with greater value and dignity (see chapter 4). Advances in moral thinking often lag behind advances in technology, and this disparity can create great turmoil, as when controversial topics like abortion become divisive and disruptive forces in society. Put another way, this fresh moral knowledge can redefine the new hybrid people so that they lose the negative aspects of their thing-like qualities and become members of Kant's shining "kingdom of ends": "... ethics views a possible kingdom of ends as a kingdom of nature. ... it is a practical Idea used to bring into existence what does not exist but can be made actual by our conduct – and indeed to bring it into existence in conformity with this Idea."[52]

COGNITIVE THINGS/COGNITIVE BEINGS

It is well known that the invention of tools and utensils at the birth of material culture initiated a sophisticated interplay between humans and things; more recently, far more complex technological products have invaded our world, ratcheting up the level of this interplay and generating a flurry of metaphors that compare humans to things and vice versa.

Tools and utensils used throughout human history were mainly an extension of men's and women's bodies; in this sense, they did not possess an independent existence and seemed to be in harmony with the environment. Modern machines, on the other hand, have independent sources of power and exist separately from the user. They have established a kind

[52] Kant, 1964, p. 104.

of third estate between nature and human arts: "While many of the boasted achievements of industrialism are merely rubbish, and while many of the goods produced by the machine are fraudulent and evanescent, its esthetic, its logic, and its factual technique remain a durable contribution: they are among man's supreme conquests."[53]

Technology represents intelligence applied systematically to the body. It constitutes a kind of prosthesis that amplifies the body and transcends its limits, compensating for its fragility and vulnerability. Because of industrial technology, the human body is capable of more than ever before, and society's unprecedented production capacity exceeds anything thought possible in the past. Usually it is said that this process has diminished the importance and diluted the talents of the worker. It is also said that technologies do not merely add something new to an environment; instead, they change the whole environment itself – ecologically, structurally, or both.[54] By anthropomorphizing things – computers, for example – we begin to devalue and disempower people while attributing too much power and value to computer technologies. From paper and pencil to computers, we have invented skillful "cognitive things." Externalizing human cognitive qualities onto machines and technological artifacts is followed by the internalizing of machines' cognitive qualities by humans. Thinking with a computer differs from thinking with paper and pencil; the computer creates a new environment in which the mind breathes a different atmosphere. Is it an information-rich world or an information-polluted world?

The new metaphors can trigger a kind of "identity crisis": "It makes a difference whether we speak of a computer as having a 'memory' or a 'data-storage capacity'. As a result of this crisis we have less sense of what is human, less sense of humans as distinct from machines, more sense of powerful machines and frail humans."[55] Things made of metal and plastic are anthropomorphized: the computer is a brain; advanced systems show artificial intelligence; machines use languages; robots have kinesthetic abilities. We endow these technologies with human traits, but we nevertheless must remember that computers and other inventions exist because a human mind imagined that they *could* exist. These technologies result when an idea for technology conceived as a "science or systematic knowledge of human arts or industrial arts" is reshaped into an idea for technology conceived exclusively in terms of objective artifacts. This transformation focuses attention on the products instead of on the producers, so that "[w]e speak of things as being *high-tech*, not people."[56]

[53] Mumford, 1961, p. 5.
[54] Strong, 2000.
[55] Gozzi, 2001, p. 148.
[56] Ibid.

In some sense, regarding technology this way renders it autonomous and abrogates its dependence on human beings. I agree with Gozzi: "This issue is not trivial, for it involves our definition of ourselves as human beings. Which is somewhat uncertain at present."[57] Human beings can seem frail and contradictory compared to the strong, stable, and reliable machines that surround us. As noted earlier in this chapter, humanoids, cyborgs, and bionic body parts complicate the issue, which I will address in more detail in the following chapter.

Qualities transferred from things to people and vice versa often carry troubling overtones: the idea of the person as a machine, which dates back to the nineteenth century, has a negative connotation, as does the phrase "organism's program." It is also said, in an ominous way, that one "deprograms" people when dissuading them from certain convictions – political or religious beliefs, for example: "Are people machines? If so, they are clearly inferior to the faster, bigger models; and they deserve to be made obsolete. Are biological processes determined by laws of chemistry and physics? If so why bother giving people all those troublesome rights and freedoms, which are illusory anyway?"[58]

Humans beings are reduced to things in all these cases, making it easier to *treat people as means*. After all, if regarded as mere machines, people do not count very much. When people are likened to such inventions, any expansion of power through technology is countered by a contraction of their self-concept, and we need more knowledge so that humans can be better "respected." Teasing out the meaning and significance of these complex issues is our duty, especially when dealing with problems in collective settings like work, school, and politics as well as in family arenas, where sex, children, relationships, and (as we have seen) reproduction can create challenges.

In chapter 3, I will explore how new knowledge can be a tool to enhance human freedom and responsibility. That discussion will lay the groundwork for understanding the idea of "knowledge as duty" in chapter 4, where I will also examine more closely our nature as "hybrid people," the fruit that results from cross-pollinating the cognitive beings and cognitive things already introduced.

The Kantian principle holds that no person should ever be thought of as just a means but should instead be considered in conjunction with an end. In this way of thinking, cloning violates the autonomy of the people produced by that technology and prevents them from living their lives according to their own desires and values. Those who adhere strictly to Kant's maxim without considering how the world has changed in the last

[57] Ibid.
[58] Ibid.

two hundred years are likely to have difficulty coping with the inevitable advances in biotechnology, and, sadly, people who come into existence through these technologies will suffer as a result. No one can stop the advance of these techniques, and there is no question that we will eventually live and work alongside cloned, genetically engineered "monsters." We have no choice but to adapt.

As is often the case, we must interrogate what we know in order to grasp what we do not yet understand. I have concluded, for example, that embryos, while not considered complete human beings, still deserve "profound respect" and "serious moral consideration"; clones, I believe, should be similarly regarded. They are not "mere" means to an end in a way that belittles or dismisses them, and for this reason I conclude that the Kantian maxim does not apply to human cloning.

We need not fear the synthesis of the human and the technological, for most people on the planet today already *are* living examples of such a blend, as we will see in the next chapter. The similarities between us and future clones are greater than critics of biotechnology suspect, and exploring these parallels will help us to fill a gap in traditional ethics – the failure to address how clones and other "new" types of people will need ethics to manage their particular problems. Put another way, by studying the nature of today's "normal people," we can begin to understand how to assign value to tomorrow's "monsters."

3

Hybrid People, Hybrid Selves

Artifacts, Consciousness, Free Will

> ... we must see the need of having nonviolent gadflies to create the kind of tension in society that will help men to rise from the dark depths of prejudice and racism to the majestic heights of understanding and brotherhood.
>
> Martin Luther King, Jr., "Letter from the Birmingham City Jail"

In a considerable part of traditional moral philosophy, machines (and things) are thought of as means having only instrumental value, whereas people are considered ends in themselves, entities with more highly regarded intrinsic value; as we have discussed, the Kantian ideal is that people should be accorded greater worth than things. But the real world, unfortunately, is not an ideal world, and current human value systems often rank some aspects of things above certain people; intrinsic value does not always trump instrumental value. Clearly, then, the old ways of thinking no longer serve us well, and they become even less relevant as modern technology continues to blur the distinction between human beings and machines and as people acquire ever more thing-like characteristics.

The prospect of people's becoming more thing-like may seem troubling, but I contend that it need not diminish the status of human beings; on the contrary, I believe it can improve conditions for many people around the world. If things are accorded great value, it behooves us to identify the thing-like qualities of human beings so that they too can enjoy such value. But how do we compare things to people? What are the relationships between the two? And perhaps most important, do we lose some of our humanness by pointing out our similarities to things and how profoundly we have become intertwined with them?

The chapter is divided into two main parts. The first analyzes in detail the fact that we are already hybrid people, as well as the fruit of a kind of co-evolution of both our brains and the common, scientific, social, and moral knowledge we ourselves have produced, at least starting from the birth of material culture thousands of years ago. I also maintain that this knowledge production has inextricably linked us with natural and artifactual externalities endowed with cognitive functions. This condition of "cyborgness" complicates the cognitive status of human beings and jeopardizes their dignity by destabilizing endowments I consider fundamentally important: consciousness, intentionality, and free will. The second part of the chapter addresses the problem of *preserving* these three important aspects of dignity; I will show how they are deeply connected to knowledge and, even more important, to the continual production of new knowledge and an ongoing commitment to modernizing ethical understanding.

HYBRID PEOPLE

For thousands of years, human beings have wrestled with the question of their "human" nature, and one way in particular that they have sought to define themselves is through comparisons to members of the animal kingdom. In traditional Judeo-Christian thought, of course, human beings are privileged over all other living things, as is clearly stated in Genesis. Nature was to be dominated, its value divinely designated as instrumental. And generally speaking, animals and plants throughout history have been viewed as resources for people – at least in Western culture – rather than as having intrinsic value of their own.

Animals have served human beings as sources of food and as beasts of burden for millennia, and while they are still eaten by most people throughout most of the world, industrialized nations, at least, now depend far less on work animals than they do on nonliving machines. As various cultures became more and more machine-dependent in the seventeenth century, a debate about what was called the "animal machine" arose: are animals mere machines? Analogously, can human beings be considered man-machines?

René Descartes's answer was that animals are, in fact, simply machines, and that human beings would be, too, if it were not for their immaterial souls. It was his well-known dualism, however, that defined people as superior to and therefore essentially different from animals. Michel de Montaigne took an opposing stance, asserting that the naturalness of beasts placed them above humans in the hierarchy of organisms. By the eighteenth century, Julien Offray de La Mettrie declared that man is an autonomous machine, no different from other mechanical beings, including animals.[1] We can say, following La Mettrie, that man and

[1] Arguing that the state of the soul depends on the state of the body, La Mettrie's view of man as a machine derives from the heuristic hypothesis that mental processes are in fact physiological. La Mettrie also introduced the critical notion that conscious and voluntary

animals are cyborgs insofar as they can be thought of as merely *material* machines; recent results in cognitive science lead us to a new and different perspective, however, as I hope to illustrate in the present chapter.

Working from Andy Clark's conclusions on the relationship between people and technology, we all are "constitutively" natural-born cyborgs – that is, biotechnologically hybrid minds.[2] It is becoming less and less appropriate to assume that our minds exist only in our heads: human beings have solved their problems of survival and reproduction by "distributing" cognitive functions to external, nonbiological sources, props, and aids. Our biological brains have delegated to external tools many activities that involve complex planning and elaborate assessments of consequences.[3] A simple example might be how the brain, when faced with multiplying large numbers, learns to act in concert with pen and paper, storing part of the process and the results outside itself. The same occurred when Greek geometers discovered new properties and theorems of geometry: using external supports (like sand or a blackboard), they manipulated external diagrams to gain important new information and heuristic suggestions.[4] Every society uses external tools, whether it is the most complex computer or a simple garden hoe; they are such integral parts of peoples' lives, the cognitive skills needed to employ them are so deeply entrenched, that such tools become almost invisible. This process gives rise to something I have called "tacit templates" of behavior that blend "internal" and "external" cognitive aspects.[5]

New technologies will facilitate this process in novel ways: on a daily basis, people are linked to nonbiological, more or less intelligent machines and tools like cell phones, laptops, and medical prosthetics. Consequently, it becomes harder and harder to say where the world stops and a person begins. Clark observes that this line between the biological self and the technological world has always been flexible and contends that this fact must be addressed by both epistemological and ontological points of view. Thus the study of the new anthropology of hybrid people becomes important, and I would add that it is also critical for us to delineate and articulate the related ethical issues. Some moral considerations are

processes are distinguished from involuntary and instinctual activities only by the relative complexity of their material and mechanical substrates. His analogy between man and machine, which went far beyond Descartes's static mechanism, led to the concept of the living machine as a purposive, autonomous, and dynamic system.

[2] Cf. his *Natural-Born Cyborgs: Minds, Technologies and the Future of Human Intelligence* (2003).

[3] Ibid., p. 5.

[4] I have devoted part of my research to analyzing the role of diagrams in mathematical thinking and geometrical discovery (Magnani, 2001b, 2002).

[5] Tacit templates of moral behavior in relation to moral mediators are treated in chapter 6. Their epistemological counterpart, which has to do with manipulative abduction, is illustrated in chapter 7.

mentioned in the last chapter of Clark's book, in which he addresses important issues such as inequality, intrusion, uncontrollability, overload, alienation, narrowing, deceit, degradation, and disembodiment – topics that are especially compelling given recent electronic and biotechnological advances. Nevertheless, Clark's approach does not shed sufficient light on basic ethical problems related to identity, responsibility, freedom, and control of one's destiny, problems that accompany technological transformations. He clearly acknowledges such issues, but only in a minimal and general way:

> Our redesigned minds will be distinguished by a better and more sensitive understanding of the self, of control, of the importance of the body, and of the systemic tentacles that bind brain, body, and technology into a single adaptive unit. This potential, I believe, far, far overweighs the attendant threats of desensitization, overload, and confusion. ... Deceit, misinformation, truth, exploration, and personal reinvention: the Internet provides for them all. As always, it is up to us, as scientists and as citizens, to guard against the worst and to create the culture and conditions to favor the best.[6]

As I contend in this book, I think these problems are more complicated, and teasing out their philosophical features will require deeper analysis. What new knowledge must we build to meet the challenges of living as hybrid people with thing-like qualities? I certainly share Clark's enthusiasm for philosophically acknowledging our status as "cyborgs," but I would like to go further, to do more than just peer through the window of his book at the many cyberartifacts that render human creatures the consumers-cyborgs we are.

Our bodies and our "selves" are materially and cognitively "extended"– enmeshed, that is, with external artifacts and objects – and this fact sets the stage for a variety of new moral questions. For example, because so many aspects of human beings are now simulated in or replaced by things in an external environment, new ontologies can be constituted – and Clark would agree with me. Pieces of information that can be carried in any physical medium are called "memes" by Richard Dawkins.[7] They can "stay" in human brains or jump from brain to brain to objects, becoming configurations of artificial things that express meaning, like words written on a blackboard or data stored on a CD, icons and diagrams in a newspaper, configurations of external things that express meaning like an obligatory route. They can also exist in natural objects endowed with informative significance – stars, for example, which offer navigational guidance. In my view, these chunks of information are externalized when

[6] Clark, 2003, p. 179 and p. 187.
[7] Dawkins, 1989.

human beings delegate them to material objects and structures, as when one jots down a phone number on the back of an envelope (cf. chapters 6 and 7).[8]

Beyond the supports of paper, telephone, and media, many human interactions are strongly mediated (and potentially recorded) through the internet. What about the concept of identity, which is so connected to the concept of freedom? At present, identity has to be considered in a broad sense: the amount of externally stored data, information, images, and texts that concerns us as individuals is enormous. This storage of information creates for each person a kind of external "data shadow" that, together with the biological body, forms a "cyborg" of both flesh and electronic data that identifies us or potentially identifies us. I contend that this complex new "information being" depicts new ontologies that in turn create new moral problems. We can no longer apply old moral rules and old-fashioned arguments to beings that are at the same time biological (concrete) and virtual, situated in a three-dimensional local space but potentially "globally omnipresent" as information packets. For instance, where we are located cybernetically is no longer simple to define, and the increase in telepresence technologies will further affect this point. It becomes clear that external, nonbiological resources contribute to our variable sense of who and what we are and what we can do.

In chapters 4 and 5, I will try to outline a new ethical framework that makes problems like identity, freedom, responsibility, and ownership of our own destinies more meaningful. For now, however, let us reconsider the consequences of my motto "respecting people as things" when applied to the idea of hybrid people.

THE BODY AND ITS CELL PHONE, THE CELL PHONE AND ITS BODY

If, as Clark holds, the line between the biological self and the technological world is flexible and continually shifting, the issues I proposed in the first two chapters appear to be further complicated. The intertwining of the biological with the technological results in many interesting "cyborgs" worth examining. Let us first consider what we can call *tool-using humans* (those who, for instance, work to procure food or shelter), beings who lived at the time when early tools like axes were devised and so-called material culture was born. Such new tools spurred cognitive development among early humans beings, and these hominids already possessed what

[8] I will address the role of this kind of cognitive delegation from an ethical perspective in the sections "Being Moral through Doing: Taking Care" (chapter 6) and "The Logical Structure of Reasons" (chapter 7) and from an epistemological perspective in "Cognitive and Epistemic Mediators" (chapter 7).

Stephen Mithen calls *different intelligences*, which comprise three cognitive domains – natural history intelligence, technical intelligence, social intelligence – each accompanied by varying levels of consciousness about the thoughts and knowledge it contains. Eventually, these isolated cognitive domains became integrated, which facilitated the development of public language: as *levels of consciousness increase*, human beings form thoughts about thoughts, and, consequently, *social intelligence* and *public language* arise.

But material culture is not just the *result* of this massive cognitive shift; it is extremely important to stress that it also *caused* the change. Mental challenges improve brain function, and using new, unfamiliar tools stimulated early human minds into more developed states: "The clever trick that humans learnt was to *disembody* their minds into the material world around them: a linguistic utterance might be considered as a disembodied thought. But such utterances last just for a few seconds. Material culture endures."[9] From this perspective, we can see that material artifacts, like written language, allow us to explore, expand, and manipulate our own minds, and the evolution of culture is therefore inextricably linked with the evolution of consciousness and thought.

The early human brain becomes an integrated, creative, "intelligent" machine so flexible that "separate" intelligent machines are not necessary for each new job. The engineering challenge to produce ever-more-specified machines for particular purposes is being replaced by the "programming" of Turing machines to do these jobs, and as a result, different intelligences are consolidated into a new universal device endowed with a high level of consciousness. From this perspective, in order for human minds to expand, they must be continuously disembodied into the *material world* around them, and the evolution of the mind is inextricably linked with the evolution of large, integrated, material cognitive systems.[10]

In addition to the tool-using humans who have existed for thousands of years, there are four other kinds of modern cyborg beings: (1) *enhanced humans*, who arise from the use of prosthetics, pacemakers, artificial organs, and so on; (2) *medical cyborgs*, who are the fruit of IVF, cloning, genetic enhancement, and so on, as mentioned in chapter 2; (3) *cognitively enhanced cyborgs*, who use devices like an abacus, laptops, cellphones, and external cognitive representations in general; and (4) *super-cyborgs*, who are endowed with stimulator implants or silicon chip transponders, for example, as I will illustrate in the following pages.

It is now clear that the biological brain's image of the body is protean and negotiable, an "outgoing construct" that changes as new technologies

[9] Cf. Mithen, 1996, p. 291.

[10] Cf. Magnani, 2006b, 2006c, and 2007.

are added to our lives.[11] Take the human visual system, for example, where much of the database is left outside the "head" and is accessed by outward-looking sensory apparatus (principally the eyes). As opportunistic cyborgs, we do not care whether information is held within the biological organism or stored in the external world in, say, a laptop or a cell phone. And not only do new technologies expand our sense of self – they can even induce changes in the actual physical body: increased finger mobility has been observed among people under twenty-five, caused by their use of electronic game controllers and text messaging on cell phones.[12]

To further analyze the externalizing of cognitive resources, let us consider a theory from the area of computer vision: the "active perception" approach.[13] This approach aims at understanding cognitive systems in terms of their environmental *situatedness*: instead of building a comprehensive inner model of its agent's surroundings, a perceptual capacity simply obtains whatever information is necessary to function in the world at that time. The agent constantly "adjusts" its vantage point, updating and refining its behavior in order to uncover a piece of information. Active perception requires the ability to examine and interpret an object effectively, to conduct detailed, methodical exploration; it is a purposeful moving through what is being examined, actively picking up information rather than passively transducing.[14] This view of perception may be applied to all sense modes: in the haptic mode, for example, mere passive touch reveals little, but actively exploring an object with our hands tells us a great deal.[15]

Biological versus Nonbiological

Clark correctly depicts a mobile phone as something that is "part of us," taken for granted, an object regarded as a kind of "prosthetic limb over which you wield full and flexible control, and on which you eventually come to automatically rely in formulating and carrying out your daily goals and projects."[16] It is well known that Heidegger distinguished between a tool's or artifact's being "ready-to-hand," like a hammer or a cell phone, and its being "present-at-hand." A ready-to-hand tool does not demand conscious reflection.

[11] Cf. Clark, 2003, p. 5.
[12] Ibid., p. 86.
[13] Thomas, 1999.
[14] Gibson, 1979.
[15] See chapter 7 for a discussion of the relationship between active vision and abductive reasoning.
[16] Clark, 2003, p. 9.

We can, in effect, 'see right through it,' concentrating only on the task (nailing the picture to the wall) [or writing an SMS message on a cell phone, we can add]. But, if things start to go wrong, we are still able to focus on the hammer [or on the cell phone], encountering it now as present-at-hand that requires our attention, that is, an object in its own right. We may inspect it, try using it in a new way, swap it for one with a smaller head, and so on.[17]

Using a tool becomes a continuous process of engagement, separation, and reengagement. Because they are "ready-to-hand," these tools are called "transparent" or "invisible" technologies.[18]

This brings me to the following point: okay, I also possess a mobile phone and have, consequently, gained a new degree of "cyborgness." I am no longer only intertwined with classic tools like hammers, books, and watches, but also "wired" to a cell phone through which I work, I live, and I think. To continue an idea from the previous section, the problem is that our enthusiasm for technological advances may blind us to the ethical aspects of these processes of engagement, separation, and reengagement.

To heighten my awareness of such processes, I, as I use my cell phone and other tools yet to come, hope to acquire the moral knowledge necessary to maintain and even reinforce my identity, freedom, responsibility, and ownership of my own future; I would hope for the same for all other hybrid people. I respect the new object or artifact that integrates its cognitive abilities with its users', but we must be mindful of the responsibilities that technology brings so that it enhances rather than diminishes us. Moreover, does the cognitive value of the artifact count more than some basic biological cognitive abilities of the human body?

Everyone has experienced the difficulty and complexity of unsubscribing from some cyberservice suppliers like cell phone companies and internet providers. Such obstacles testify to the fact that even if they are effective tool-based cognitive extensions of our bodies, they also are tool-based economic institutions aiming to cast themselves as cognitively necessary and irreplaceable things. Because they satisfy market needs, they in some sense acquire more importance than the biological life itself.

As illustrated earlier, new artifacts become "ready-to-hand," but at what ethical cost? We must still be able to extricate ourselves, if we so choose, from the technology that has appeared in our lives. Terminating a cell phone service contract, for example, should be an easy process without extended hassles or unexpected costs.[19] What way of ethical thinking fully explicates that right and will lead to new policies and laws

[17] Ibid., p. 48.

[18] Weiser, 1991. On the so-called invisible technologies, see Norman, 1999.

[19] A human being may feel that while all people are mortal, one's subscription to the internet provider will never die. The "lives" of these small external artificial things tend to overcome our own.

that will protect human dignity in the future? What moral knowledge will I need if a sophisticated new neurophone[20] is wired into my cochlear nerve as a direct electronic channel? Or how will you get rid of an "affective wearable" that monitors your stress level and provides daily profiles and other data to you, but in the meantime is generating an intolerable information overload?[21] You start to think you have another "self," and it feels as if you no longer own some of the information about yourself – that damn "affective wearable" also monitors all your frustrations and shows you an interpretive narrative on how things are going. Managing this alien, technologically induced hybrid self can be difficult; the new attributes we gain through technology can leave us feeling awkward and ungainly, and it is only through knowledge that we can learn to navigate in our "new" bodies. This knowledge needs to be established sooner rather than later: advances like the neurophone Clark describes and the "affective wearable" will exist before the moral and legal rules that apply to them are worked out.

As I will illustrate more fully in the following chapter, appropriate knowledge will help to protect us from the dangerous ethical consequences of future technology. As Brad Allenby remarks, "It is usual to deal with technology and technology systems at many scales, but it is perhaps too seldom that we stand back and evaluate what humans, as toolmakers, have really created: the anthropogenic Earth, a planet where technology systems and their associated cultural, economic – and, indeed, theological – dimensions increasingly determine the evolution of both human and natural systems."[22] Human activity now has a profound impact on the evolution of the planet, an impact that grows greater with each passing year through both ecological consequences and biotechnological practices.[23] Allenby maintains that we must use rational and ethical approaches to continually manage human-natural systems in a highly integrated fashion – an Earth Systems Engineering and Management (ESEM) capability. It is from this perspective that I will delineate the new moral responsibilities we must accept if we wish to limit the damage we inflict on the planet and on ourselves.

While we are familiar and comfortable with heart pacemakers and simple cochlear implants, on the horizon are new bionic devices that will invade the human body in unprecedented ways. Kevin Warwick predicts new super-cyborgs formed by human (or animal) machine brain/nervous system coupling.[24] There are stimulator implants that electronically

[20] Clark, 2003, chapter 1.
[21] Picard, 1997, p. 236.
[22] Allenby, 2005, p. 303.
[23] Cf. the previous chapter.
[24] Warwick, 2003.

counteract the tremors associated with Parkinson's disease; implants that, by sending signals from the brains of stroke victims to computers, will allow patients to communicate by forming words on a screen; and silicon chip transponders to be surgically implanted in the upper left arm that will transmit unique identifying radio signals. Other chips, once linked to nerve fibers in the arm, transmit and receive signals that trigger movement; similarly, researchers have been able to transmit signals directly from a human brain implant across the internet to a robot, inducing it to raise its hand. And more surprising still, extrasensory inputs gathered by robots' ultrasonic sensors have been sent to human brains that, amazingly, have succeeded in making sense of the signals.

Of course, as Warwick and Clark claim, these super-cyborgs can control evolution so that it is based entirely on technology rather than biology. Even more than it already is, evolution will become co-evolution: the organism (already hybrid) and the continuously modified environment (in turn disseminated by super-cyborgs) will find a continuous mutual variation. Organisms continually adapt to their surroundings to ensure survival, but at the same time, the environment is also constantly transforming. Within this complex system of changes, many organisms might fit into the same niches, and the ecosystem that contains them becomes highly sensitive and reactive to the organisms that live in it.

Advanced technologies involving surgical manipulation of the brain have led to the new field of neuroethics, which also addresses problems related to the possibility of the super-cyborgs I have just illustrated.[25] While surveying the history of electrical brain stimulation and its coalescence into the present-day neuromodulation and psychosurgery debate, Fins[26] quotes an editorial from the *Economist* of May 2002 that anticipates greater "threat[s] to human dignity and autonomy than cloning" will arise from advances in neurology and psychiatry. It is clear that these developments further blur the line between mind and brain, a distinction upon which much of our diagnostic typology depends.[27]

It is not only the cell phone (or the affective wearable), which is now so wired to human bodies, that has quickly acquired cognitive dignity and a subsequent positive moral status. As we have said in the previous chapters, many "things" ("means") that are initially devoid of any kind of value or have been assigned only a "market price" or "affective price" can also acquire a moral status (an intrinsic value, it is said). This

[25] Safire, 2005.

[26] Fins, 2004.

[27] In the case of psychosurgery, the problem of gaining consent from people who lack decision-making capacity is still an open and relatively unexplored ethical issue. The issue has also been related – especially in the seventies – to the political debate on possible mind control and racial repression and, of course, to the problem of free will and determinism (see the discussion later in this chapter).

transformation has occurred for some natural "things" like women and animals as well as for artifacts like important paintings, so it is not strange that everyday artifacts that are cognitively intertwined with people – mobile phones, laptops – would achieve a similar status.

These kinds of "things" acquire respect because of their ability to increase human cognitive faculties, to improve people's ability to communicate, act, and reason. These "things" are able, in some sense, to "augment reality." "Augmented reality" was the term used in the more sophisticated case of a group of Boeing engineers and scientists in the early 1990s who sought to construct complex wiring harnesses in aircraft so that workers would see the desired positioning superimposed upon the actual physical structure of the plane.[28] The augmented information could also be picked up by using special eyeglasses or other wearable devices.

I have already explained in chapter 1 that, following Kant's anthropocentric ideas, "things" have had to struggle, so to say, to gain worth and value; such is also the case for human beings who, being almost completely deprived of human dignity, just wish to stop being mere "means."

There is a profound tension between the biological and the technological spheres of human hybrids, who are composed of a body plus cell phone, laptop, the internet, and so on. Sometimes the two aspects can be reconciled by adjusting and redistributing various values, but the struggle is ongoing, and the final results are unknowable: the outcome simply depends on the moral targets hybrid people identify and advocate. Does the cognitive value of the cell phone count more than some cognitive value of the biological body? Is this delegation of tasks to the cell phone really compensated for by new capabilities, or does a biological body's lack of cognitive autonomy become intolerable at some point?

Human beings are able to "actively" invent many ways to dispossess themselves and others (cf. chapter 1). This also happens when new scientific knowledge with the potential for unforeseen technological outcomes is employed in certain social settings and ultimately causes harm to various kinds and categories of people. Only later do victims of such behavior tend to reclaim their ambitions and aspirations and to arrive at (or maintain) a full and flourishing life.

The economic value of technological objects that are "grafted" onto human beings makes it dangerously easy to treat people as means, and it is well known that the market economy is inherently inclined to regard human beings in this way. In a market economy, the qualities and worth of human beings – their intelligence, energies, work, emotions, and so on – can be "arbitrarily" exploited and/or disregarded in favor of promoting

[28] Clark, 2003, p. 52.

the sales of items that may or may not be particularly useful. Such situations, of course, inevitably generate frustration among those whose interests are pushed aside.

Central to this issue is the fact that many people are used to being considered things: they are, in Kantian terms, "treated as means (and only as means)." In chapter 1, I offered a way to recalibrate the value of things so that "respecting people as things" becomes a positive way to regard human beings. To give an example, imagine people who have used certain devices so much that some of their biological cognitive abilities have atrophied. Such people may yearn to be as respected as a cell phone – perhaps the expensive one of the future that I mentioned earlier, the direct electronic channel wired into my cochlear nerve that features a sophisticated processor, spectacular AI tools, and a direct internet connection. The hybrid person of our example will feel herself dispossessed of the moral cognitive worth attributed to nonbiological artifacts. It is very easy to imagine how this situation will be increasingly complicated by the appearance of future super-cyborgs endowed with huge extra memory, enhanced mathematical skills, extrasensory devices, and – why not? – the ability to communicate via thought using various signals. They will be more powerful than garden variety hybrid people with brains that are part organic tissue and part machine, so that the "epicentre of moral and ethical decision making will no longer be of purely human form, but rather it is a mixed human, machine base."[29]

As already discussed, being cared for and valued is not always considered a human right; collectives, for instance, do not have moral (or legal) rules that mandate the protection and preservation of human beings' cognitive skills. As a result, we face a paradoxical situation that inverts Kant's thinking, one involving people who are not treated as well as means or things are treated. Yet people's biological cognitive skills deserve to be valued at least as much as a cell phone: human cognitive capacities warrant moral credit because it is thanks to them that things like cell phones were invented and built to begin with. In this way, human hybrids can reclaim "moral" recognition for being biological carriers of information, knowledge, know-how, autonomy, cultural traditions, and so on, and gain the respect given to cognitive artifacts that serve as external repositories: books, for example, PCs, or works of art. The human hybrid who exhibits the knowledge and capacity to reason and work can expect to play a clear, autonomous, and morally recognized role at the level of his or her biological intellectual capacities.

We can say that the hybrid character of human beings has also made it possible to attribute ethical importance to many artifacts, especially those we regularly keep close to our bodies. This is true of cell phones and other

[29] Warwick, 2003, p. 136.

technological tools, like laptops, medical devices, and electrical power that have, step by step, acquired value beyond their obvious utilitarian, economical, and social value; they have gained their own intrinsic moral worth. This is the reason I maintain that these artifacts, which are already endowed with the intrinsic moral value we have collectively assigned to them, can be considered *moral mediators*. As already illustrated, moral mediators are distributed throughout in the environment and extend value to human beings, to other nonhuman things, even to "non-things" like future people and animals (I will discuss *moral mediators* in detail in chapter 6). We are surrounded by that which is human-made and artificial, not only concrete objects like a hammer or a PC, but also human organizations, institutions, and societies; all of these things have the potential to serve as moral mediators. For this reason, I say again that it is critically important for current ethics to address the relationships not only among human beings, but also between human and nonhuman entities.

With slight modifications, what I have just said also applies to the previously mentioned super-cyborgs. Two moral problems are still at stake: (1) the difficulty of distributing sophisticated technologies like extrasensory devices equally among the world's people,[30] and (2) the risk of overvaluing super-cyborgs' biotechnological cognitive skills: we must not privilege their artificial cognitive capacities over their organic cognitive mechanisms and thus have beings with powerful prosthetic intelligence devices but dull, "natural" brains.

ETHICS OF SCIENCE VERSUS ETHICS OF KNOWLEDGE

In the previous section, I declared my hope for acquiring knowledge that will allow us to maintain and enhance our freedom, understand our responsibility, and preserve ownership of our futures given the current onslaught of technological advances. While I respect new objects or artifacts that integrate my cognitive activities, I believe it is imperative that we explore the moral implications of such devices before embracing their use.

My point is clear: I care about the cognitive capacities of present human beings, and I also admire their capacity to be conscious and intentional agents with the benefit of free will and the ability to take responsibility for their actions. As will become clear in the following sections of this chapter, I naturalistically consider free will, like the mind,

[30] It is evident that current human brains are already provided to varying degrees with external "natural" (teachers, parents, other human beings, etc.) and artificial (books, schools, laptops, internet access, etc.) cognitive mediators, because of biological differences and social inequalities. For more information about cognitive delegations to organizations, institutions, and so on, see Perkins, 2003.

to be an evolutionary product enjoyed by present human beings. Free will is not a preexisting and ontological human feature: like human dignity, it can be weakened or lost, possibly giving rise to less appealing – at least for me – anthropologically new creatures.

Indeed, aspects of human dignity are constantly jeopardized by human wrongdoing, mistakes, bad politics, misery, and so on, as well as by technological products, as I am illustrating in this book. Constant challenges come also from natural events and transformations (both normal and extraordinary, like epidemics and catastrophes). I think that maintaining and improving current human characteristics depends on our choices about knowledge and morality. The pursuit of scientific knowledge allows us to better understand the external events of nature and to develop technological artifacts that protect us and provide what we need. Nevertheless, as I have tried to illustrate in the previous chapters, we are rarely fully aware that external "things" like technological tools can jeopardize our dignity, if we agree that "dignity" comprises the aspects described earlier – free will, freedom, and responsibility.

I have already declared that I think modern technology has brought about consequences of such magnitude that older ethical frameworks and policies can no longer contain them. For example, modern technology has turned nature into a human responsibility, and as a result, we must now approach it not only with unprecedented cleverness but also with new ethical knowledge. As I will explain in detail in the following chapter, I maintain that knowledge has become a much more profoundly important *duty* than ever before, and this knowledge must be commensurate with the causal scale of our action. Today, it seems only scientific and technological advances earn society's esteem, but information generated *only* by science and technology creates a lopsided wisdom that is blind to future conflicts. The answer is to balance technological achievements with commensurate accomplishments in ethics.

The Intrinsic Limitations of the So-Called Ethics of Science

To examine this new commitment to knowledge, it is important to consider the intrinsic limitations of the so-called ethics of science. As I described earlier, preserving and enhancing some ethical aspects of human beings depends on their own choices – that is, human dignity depends on a kind of general "ethics of knowledge" that must be addressed before any "ethics of science" is specifically discussed.

A new discipline that has appeared only recently, the ethics of science explores moral problems faced by people involved with laboratories and scientific institutions. The field must be distinguished from the ethics of technology, which examines moral aspects of the relationships that technological artifacts themselves have with both nature and human beings. Many books and articles are already available on the ethics of science,[31] and they address a variety of subjects in compelling ways. Science is a profession with a vast number of issues to address: plagiarism, fraud, violations of the law, mismanagement of funds, mutual respect among scientists, violations of recombinant DNA regulations, discrimination, conflict of interest, and the exploitation of subordinates (the mentor-mentee relationship, harassment, reporting misconduct, hiring and recruitment, sharing and preserving resources). Despite a growing body of evidence on unethical research, however, the data indicate that the frequency of scientific misconduct is low compared to that in other professions such as business, medicine, and law.[32]

Some of the moral problems we find in scientific research plague other professions as well, but they do not require new, specialized moral frameworks because those fields are both comparatively easy to understand and familiar to most people. By contrast, myriad ethical ambiguities are peculiar to science, problems generated by the complexity of scientific institutions and the work they undertake, and these challenges require more sophisticated rules and principles, and even entirely new moral constructs. Areas that require such detailed attention include the following: social responsibility, peer review, respect for both human and animal subjects, resource allocation, conflicts that occur when scientists' personal and financial interests clash with their professional obligations, bias in research, ownership of intellectual property, giving credit where credit is due, data management, scientists' role in industry, providing expert testimony in court, military science, ties between the individual and the collective, internal and external ethical aspects of various situations, knowledge deficiencies and moral responsibility, laboratory risks, privacy, reprisals for withholding or for spreading truth, authorship, and environmental concerns. With all these issues on the table, the global scientific community agrees that ethics committees must be established at every research institution and that ethical standards for scientists must be enhanced or, in some cases, resurrected and applied with greater urgency.

Discussing the intrinsic limitations of the ethics of science is worthwhile because it will help me to develop a compelling new ethical perspective on knowledge. The ethics of science is too narrow to help us interrogate the concept of hybrid people: it is a broad "ethics of knowledge" that can help us to manage societal changes wrought by technology. The commitment to acquire such knowledge is conspicuously absent from the ethics of science, and it seems the concept is either not acknowledged at

[31] For recent examples, see Resnik, 1998; Kitcher, 2001; and Seebauer and Barry, 2001.
[32] Resnik, 1998, p. 1.

all or deemed irrelevant to ethical consideration because the field primarily involves only scientists and research funds. I hope to prove, however, that there is in fact a profoundly important connection between broad-based ethical knowledge and science, and to that end, I have compiled numerous examples of such links and here offer observations on why we are obliged to understand them if we truly wish to increase human dignity.

As I will more fully outline in the following chapter, the rise of modern science has generated an increase in *rational* knowledge. In the last century, especially in Western Europe and the United States, the fields of philosophy, logic, epistemology, and cognitive science have helped to sketch a wonderfully enlightening picture of the structure of both technological products and human rational capacity. But comparable strides have not been made in at least three critical areas – in developing general ethics, in exploring ethical issues surrounding public policy, and in studying the interplay among rationality, technology, and *values*. I strongly believe that we are far from successfully constructing a modern, *self-correcting* body of ethical knowledge; what we have now is severely inadequate given the present and future need to manage relationships between the poor and the rich in national settings and in the global arena.

This knowledge gap has made it impossible to protect or respect the human characteristics I care about – consciousness, intentionality, free will, freedom, and responsibility; we have neglected these aspects in the same way that we have failed to interrogate critical environmental and biotechnological issues, as we saw in chapters 1 and 2. I have also said that I think all these aspects are constantly jeopardized by human beings themselves, both directly (simple immorality, wrongdoing, politics, etc.) and indirectly through the technological products themselves.

Maintaining and improving certain aspects of the lives of people around the world depends on our own choices about knowledge and morality – that is, on a kind of "ethics of knowledge" that must be established beyond a useful "ethics of science." This ethics of knowledge informs science and serves as its framework, and we must build a flexible, self-correcting body of philosophical and ethical knowledge that can see us through the technological changes to come. In keeping with this line of thought, I offer in the following sections a philosophical and cognitive illustration of how consciousness, intentionality, and free will are linked to the production of scientific and ethical knowledge. Understanding this connection will allow us to explore more deeply the ethical issues of freedom, responsibility, and the ownership of our own destinies that I will address in chapters 4, 5, and 6.

CRITICAL LINKS: CONSCIOUSNESS, FREE WILL, AND KNOWLEDGE IN THE HYBRID HUMAN

Consciousness, intentionality, and free will – each one appears to me a vitally important endowment, whether we speak of ourselves as "traditional" human beings or as hybrid human beings. The problem, however, is that it is difficult to preserve and enhance these qualities for cyborgs because of their unfamiliar nature and greater level of complexity. How can we understand hybrid people so that they too have equal access to such endowments? Once again, I believe the answer is *knowledge*.

As we have established, the hybrid person consists of biological parts as well as internal and external representations, of internal tools and mechanisms as well as external devices that serve cognitive and moral functions. I submit that, like people, consciousness and free will can also be considered hybrids, "products" that result from the merging of one's *internal* knowledge and ethical commitment with *external* knowledge and ethical commitment. Human consciousness and free will are shaped not only by factors that are "inside" the hybrid, but also by those located in the "outside" world, where they can be picked up by hybrid agents' senses.

Let us take a look at these problems from a neurocognitive and psychophilosophical perspective. Consciousness is a staggeringly complex state: it includes the "perceptual world; inner speech and visual imagery; the fleeting present and its fading traces in immediate memory; bodily feelings like pleasure, pain, and excitement; surges of feeling and of emotion; autobiographical memories as they are recalled; clear and immediate intentions, expectations, and actions; explicit beliefs about oneself and the world; and concepts that are abstract but focal."[33] Neuroscientists who study consciousness always say that their science should also address the so-called hard problems posed by such concepts as voluntary action, free will, qualia, sense of self, and the evolution of consciousness.[34]

Voluntary Action and Free Will

Stanislas Dehaene and Lionel Naccache see consciousness as a unified "neural workspace"[35] through which many processes can communicate. It can be hypothesized that at the level of the brain

> ... the hypothesis of an attentional control of behavior by supervisory circuits including AC and PFC, above and beyond other more automatized sensorimotor pathways, may ultimately provide a neural substrate for the concepts of voluntary

[33] Baars, 1997, p. 3.
[34] Dehaene and Naccache, 2001.
[35] A metaphor already used by Baars, 1997.

action and free will (Posner, 1994). ... One may hypothesize that subjects label an action or a decision as "voluntary" whenever its onset and realization are controlled by higher-level circuitry and are therefore easily modified or withheld, and as "automatic" or "involuntary" if it involves a more direct or hardwired command pathway (Passingham, 1993).[36]

Dehaene and Naccache further observe that human beings often make voluntary decisions by setting a goal and then selecting a course of action by serially examining many alternatives and evaluating their possible outcomes. This conscious process corresponds to what subjects refer to as "exercising one's free will"; it could be characterized as a kind of

... decision-making algorithm and is therefore a property that applies at the cognitive or systems level, not at the neural or implementation level. This approach may begin to address the old philosophical issue of free will and determinism. Under our interpretation, a physical system whose successive states unfold according to a deterministic rule can still be described as having free will, if it is able to represent a goal and to estimate the outcomes of its actions before initiating them.[37]

The authors clearly conciliate determinism with free will. This perspective, called "compatibilism" in the literature, holds that free will *is* compatible with determinism, that choices exist even in the face of predetermined conditions a being had no role in creating. To say that an action is determined means that it, like any other event, results from some cause or causes; to label an action as free simply means that it is determined by certain kinds of causes and not others. I may choose to pass my summer holidays at the seaside because, for me, swimming there generates pleasurable emotions, but the fact that my decision is emotionally driven does not render it less free.

Daniel Dennett places compatibilism and the emergence of free will in an evolutionary context:[38] determinism, he contends, does not imply that our natures are fixed and that whatever we do, we could not have done otherwise. Rather, he argues that natural selection offers progressively greater degrees of freedom even for relatively simple organisms, not to mention for people, and that the emergence of human free will is perfectly compatible with evolutionary theory and with any deterministic or indeterministic scientific description of reality. The opposite view is called "incompatibilism" and maintains that if determinism is true, free will does not exist.

[36] Ibid., p. 29.
[37] Ibid., pp. 29–30.
[38] Dennett, 2003.

Philosophers like such problems, and the literature on these themes is vast, even if it is sometimes very speculative. I just want to add that the second view – incompatibilism – obviously favors the traditional dualistic notion of a division between mind and body, an idea called "interventionism." To overcome this dualism in incompatibilism, some philosophers (called "libertarian") in turn stress that the quantum account also guarantees randomness and indeterminism at the level of the brain. Two problems immediately arise: (1) the conscious capacity of free choice is not supposed to inherit the randomness of quantum mechanics; and (2) if it were, "the indeterministic spark occurring at the moment we make our most important decisions couldn't make us ... more self-made or more self-autonomous in any way that could be discerned from inside or outside, so why should it matter to us?"[39] Moreover, brain cells should serve to stabilize quantum indeterminacy at the subcellular level: decision making is supposed to be causal, and it would be unsettling to learn that our brains arrive at decisions in the same random way that a ball settles into a numbered slot on a gambler's roulette wheel.[40] For other researchers, the indeterminacy effect would be the same whether we were trying to identify an object in a rapid sequence of different perceptions or internally scanning through memorized patterns (motivators) in order to decide about an action.[41]

Evolution of Consciousness

In the course of phylogenesis, the emergence of consciousness can clearly be associated with the increased independence it guarantees to human beings. The ability to mentally simulate and evaluate different courses of action seems much more advantageous for an agent than simply plunging into action without regard for context or precedent, which would be both risky and highly energy-consuming: "By allowing more sources of knowledge to bear on this internal decision process, the neural workspace may represent an additional step in a general trend towards an increasing internalization of representations in the course of evolution,

[39] Ibid., p. 36.

[40] An updated and clear explanation of the relationship between compatibilism and libertarianism is given in Searle, 2001 (chapter 9), and Dennett, 2003 (especially chapters 1 through 5). In two later sections, "Tracking the External World" and "Tracking Human Behavior," I will illustrate the basis of human free will and freedom in something more manifest than quantum indeterminacy.

[41] For an interesting analysis of quantum aspects of chaotic neurons dynamics, see Arecchi, 2003. On the relevance of neuronal quantum aspects of free will, see Stapp, 2001. An elegant quantum field theory of brain operation has been developed by Vitiello (2001). The debate on "physical reality and consciousness" is illustrated in Hameroff, Kaszniak, and Chalmers, 1999, part VII.

whose main advantage is the freeing of the organism from its immediate environment."[42]

Consciousness also comprises diverse epigenetic factors: each workspace state is highly differentiated, and although the major organization of this repertoire is shared by all members of a species, its specific contents accrue epigenetically and are therefore specific to each individual. Thus the contents of perceptual awareness are complex, dynamic, multifaceted neural states that cannot be memorized or transmitted to others in their entirety. These biological properties seem potentially capable of substantiating philosopher's intuitions about qualia of conscious experience.[43]

The previous "neuropicture" can help us to explore more deeply the nature of consciousness and, consequently, to understand more fully the ethics of freedom, of responsibility, and of owning our destinies, ideas that I will address in chapter 4 (in connection with the problem of identity and privacy) and in chapter 5 (in connection with the problem of bad faith and other philosophical issues concerning equality and globalization).

If we adopt Dehaene's and Naccache's perspective on consciousness and its evolution, it follows that, in order to perform with intentionality and free will, our brains (at the minimum) need

1. a decision-making mechanism (an "algorithm," to use a computational term); and
2. representations of possible goals and of potential actions to reach the chosen goal.

We know that both of these cognitive endowments are available to human beings: if not already stored in memory, mechanisms and representations can be learned – from Mom, for instance, or from reading a book and adapting its contents to an individual's needs. For example, I use certain "representations" when choosing a newspaper for my early morning updates on worldwide political wrongdoings. I choose from a list of possible newspapers with certain known characteristics: perhaps I already have that list in my memory, but if not, I just gather the information from somewhere "out there" in the external material world.

The skills I need for decision making are very common and easy to learn. Humankind developed this kind of reasoning long ago and, over time, learned to represent it externally in language so that it is communicable and learnable. More recently, knowledge engineers and AI scientists have constructed "algorithms" that, when worked into programs and loaded onto a PC (that is, a nonhuman thing), can perform that kind of

[42] Dehaene and Naccache, 2001, p. 31.
[43] Ibid., p. 30.

reasoning. As I have already said and will more clearly illustrate in the following sections, human hybrids' consciousness and free will are possible because there is something "inside" *and* because there is something "outside," "over there," in the world, that those hybrid agents can pick up. Evolution has given human beings the ability to use internal as well as external resources when faced with a decision, and I agree with Daniel Dennett that developing these resources greatly enhances the potential for human consciousness and free will.

TRACKING THE EXTERNAL WORLD

Tracking the External World through Everyday Knowledge

How can we identify and employ the many mechanisms of choice available to us in a way that renders our free will even more "free"? I think that everyday, philosophical, and scientific knowledge about natural and artificial phenomena (as well as the technologies that relate to them) has allowed human beings a wide range of possibilities for choosing and acting: in many ways, the more one knows, the more options one has.

The evolution of knowledge and its externalization in objects and artifacts is directly related to our bodies' and brains' capacity for consciousness and free will. Indeed, knowledge sheds light on both natural and artificial external worlds, revealing two kinds of phenomena: (1) regular, foreseeable phenomena whose predictability makes possible a range of free choices and allows us to plan appropriate and effective responses and (2) unpredictable or unprecedented phenomena that cannot be altered by human intervention because we lack the knowledge to do so. Moreover, as we well know, this second sort of knowledge, because of its dynamics, does not involve "ontological" limitations: a phenomenon once considered beyond the reach of human action can, with new understanding, become easily manageable, a process that increases our options and, consequently, our free will. In the knowledge framework of some primitive people, for example, the course of a river was not considered modifiable; but new knowledge and new artifacts made it possible to govern flowing water, allowing other ancient civilizations to "choose" between natural and artificial courses.

Dennett lists five different kinds of natural phenomena that relate to the "elbow room" he claims is the basic requirement for free will: those that are fixed, beneath notice, changing (and worth caring about), trackable ("at least under some conditions – and hence efficiently and usefully predictable under those conditions"), and chaotic – that is, (practically) unpredictable but still worth caring about.[44] These categories of phenomena – which,

[44] Dennett, 1984, p. 109. Dennett also sees chaotic systems as "the source of the 'practical' (but one might say infinitely practical) independence of things that shuffles the world and makes it a place of continual opportunity" (1984, p. 152).

I repeat, are dynamic products of human knowledge – provide us with what Dennett calls "epistemic possibilities," which are necessary in order for an agent to become a free deliberator. This variety encompasses everything that is "possible-for-all-the-deliberator-knows-or-cares" and ensures that every "deliberator-agent – a species, for instance – will always be equipped with a somewhat idiosyncratic way of gathering and partitioning information about its world so it can act effectively in it. ... It is this epistemic openness, this possibility-for-all-one-knows, that provides the elbow room required for deliberation." This epistemic openness also nullifies deterministic objections to free will.[45]

Dennett notes, however, that there is a drawback to having a great deal of space for moral navigation, and he alludes to a hazard faced by human beings who constantly create their own elbow room, even if they do manage to avoid being "too fully understood": "Just as the pilot's meta-level control planning leads him away from situations in which he risks a diminution of control, so our meta-level control thinking may lead us to want to eschew tactics of control strategies that run the risk of being too fully understood, and hence anticipated, by a competitor (if ever one should appear)."[46]

Two simple examples will clarify the problem of choice and its "elbow room" – one related to everyday situations, the other related to science (cf. the following section). It is supposed that I have a normal brain endowed with normal consciousness, like the one I described earlier with the help of Dehaene's and Naccache's scientific hypotheses. Phylogenesis and (my personal) epigenesis,[47] have provided me with "hardware," the physical equipment that can be described as having free will because it can adopt various cognitive approaches to decision making, strategies human beings are now also able to model through algorithms (in my case, the nontechnical ones I have acquired and refined, especially during my childhood).

At this moment, I am in New York City writing on my laptop; later today, I plan to go to Riverside Park, which seems to be very close. My everyday knowledge of Manhattan (some of it already stored in my memory) tells me that I can "choose" many routes to reach the Hudson River. I already have the mental representation of a possible route I have used in the past, but I want to find a new one that is much shorter. All I need are new representations of the streets, which I can easily "pick up" from a map. Most of the world's information is stored in external "mediators"– it is not necessarily found only in brains![48]

[45] Dennett, 1984, pp. 111 and 113.

[46] Ibid., p. 66.

[47] Cf. the earlier consideration of the enormous combinatorial diversity of consciousness.

[48] The idea of information as chunks of "memes" that can jump from brain to brain and from brain to external objects was described earlier in the section "Hybrid People."

At this point, all the required elements for decision making are present: brain, consciousness, intentionality, free will, a reasoning mechanism guided by a system of values, and adequate knowledge about my goal and the possible routes I might take. Some aspects are internal, and some are external. Some external ones – new data about streets and possible routes, for example – become internal when they are put in memory and are therefore represented in my brain.[49] I know there are routes I cannot choose because they involve loops or a cul de sac. At this point, I can plan my route.

At this stage, I can do more: I can externalize some data that are internal by drawing a simple map for the brain of a friend who needs information about how to get to the park. To an observer of this hypothetical interaction – a third brain – all aspects of this exchange between my friend and me are "external," and even my brain is considered an external object by the observer. But that third brain easily hypothesizes that my drawings derive from my existing "internal" representations. Some cognitive psychologists contend that human consciousness requires a capacity to infer others' mental states from their behavior.[50] In turn, philosophers use those hypothesized mental states as evidence that being aware of our own internal representations derives from the immaterial Cartesian *cogito*, that famous supplemental and embarrassing "ghost in the machine."

The recent discovery of "mirror neurons" in the premotor cortex of monkeys has provided evidence of specific neurological responses to intentional, goal-directed motor actions.[51] Researchers studied particular brain cells that fire only when a monkey performs a very specific behavior – breaking open peanuts, say. Interestingly, they discovered that the cells also fire when a monkey merely *sees* another monkey cracking peanuts.

(Information as stored in what I call cognitive and epistemic mediators – repositories of external representations – is described in chapter 7 of this book.)

49 When I arrived in New York the first time, I had to "pick up" almost all detailed knowledge from external representations (street and subway maps) and from the internal representations – externalized to me – of New Yorkers' brain/mind systems. This process built the first elements of a database in my brain/mind system for "possible" free decisions about routes in Manhattan.

50 Wegner, 2002.

51 Cf. Rizzolatti et al., 1988; Gallese et al., 1996; and Rizzolatti and Arbib, 1998. Iacoboni (2003) demonstrates how intention can be understood through imitation, a perspective that sheds light on the formation of complex social cognitive skills like intentions, planned action, empathy, mind reading, and the emergence of language. The embodied and prereflexive effect carried out by the sensorimotor system also seems to be particularly important in bodily recognition and emotion sharing (for instance, the experience of painful sensations). It is also supposed to be the foundation of empathy, because the embodied process of pretension is both a-centered and, at the same time, the basis of self-other distinction (on these problems, cf. Gallese, 2006).

The nature of these mirror neurons seems to support the hypothesis that there is an innate connection between the self and the other. In this case, bodies play the role of quick-response devices that enable us to "act" and perhaps even "feel" by seeing others' experiences. Certain manipulations of objects and gestures, then, would be innately communicative.

Let us come back to the problems of conscious free will and the role of knowledge, both internal and external. The simple example of my quest to reach Riverside Park has a clear epistemological significance; the new knowledge involved, both available and represented, indicates that based on the external mixture of natural things and artifacts that is Manhattan, there are various routes shorter than the one already stored in my mind. These shortcuts are then revealed and evaluated by the decision-making mechanisms operating in my brain. By picking up new knowledge from external devices (a map, for example) and re-representing it in my brain cells, I can enhance my range of possible choices and so gain more room to flex my free will.

As we can easily see, all the "characters" in this theater of consciousness are intertwined with one another.[52] Consciousness in human brains (and so "my" consciousness) has evolved in this way because human brains in turn have produced knowledge about the world that sanctioned as available and believable multiple representations that, once re-represented in the brain, further expand our menu of possible choices, and vice versa. Consequently, I contend that consciousness and the higher mechanism of knowledge are very much interdependent.

The only character I no longer personally see as part of the play in this theater is the Cartesian *cogito*, because I feel that my internal "free will" representations are completely material, just like the nice plastic map of Manhattan! Of course, I also think they express something voluntary and mine, of my *cogito*, but I know that this is the rough and spontaneous "explanation" that the "I" expressed by a physical mechanism in my brain gives to itself of those other physical mechanisms involved in free will performances that still occur only "over there," in my (merely material!) brain.

Tracking the External World through Philosophical and Scientific Knowledge

It is not only everyday knowledge that helps us to track the external world; higher levels of knowledge, like that found in philosophy and science, multiply the options available to human beings for choosing and acting. Consider for a moment a set of external natural circumstances and/or a lack of knowledge that renders voluntary choice impossible – a

[52] Baars, 1997.

volcanic eruption, for example. Stored in my brain, which is endowed with consciousness, is enough general knowledge of volcanology for me to understand that neither I nor any other human being can possibly predict the exact date of the next eruption of Vesuvius; even scientists who are experts in the field can only estimate that it will probably occur some time in the next two hundred years. Consequently, in the next two hundred years I cannot freely choose when to go to Naples and be certain that I will avoid an eruption. Human beings currently have a limited ability to predict and thus to affect the impact of geological events in particular and of chaotic events in general – we cannot, therefore, reconfigure the Naples soccer team's schedule so that a game doesn't coincide with the eruption. In this case, our capacity for free will is of no use: even with help from scientific models, computational devices, and complicated calculations, our brains are not able to pinpoint the precise day Vesuvius will erupt. Our best choice might be to postpone a trip to Naples until after 2206! Future improvements in volcanology, however, could allow more accurate predictions, even if we know that unpredictability is constitutive of chaotic phenomena.

If the world "out there" were always cognitively dark and homogeneous, free will would be impossible: free will and knowledge are two sides of the same coin. In recent human evolution, there has been a general increase in the human production of philosophical and, subsequently, scientific knowledge (and in their systematic "externalization") as people have sought to free the human organism from its immediate environment. It is easy to imagine how many representations and inferential mechanisms can be stored in our brains and used at will: even those not already consigned to our brain memory exist "over there," crystallized in various external mediators throughout the history of civilization, ready to be "picked up" when needed. To that same end, human beings also participate in the reverse process by externalizing many techniques and technologies, as we shall see later in this book.

The roles of knowledge, however, extend beyond the phenomena of the natural and artificial world, beyond settings like the streets of a city. In the external world there are also other human beings. What happens to our internal free will mechanisms when we are faced with the behavior of other human beings?

TRACKING HUMAN BEHAVIOR

Rendering Human Behavior Predictable through Ethics

Humankind has developed consciousness and compiled everyday and scientific knowledge about the phenomena of the natural and artificial world in a way that has yielded "better brains" (and thus higher levels of consciousness) and has, in turn, made higher levels of knowledge possible. So consciousness, with help from its endowments of intentionality and free will, has nourished, produced, and "externalized" knowledge and given rise to modern science. How can we think of the birth of modern science without considering the specialization in human beings of the role of voluntary decisions, intentionality, free will, responsibility, planning, and so on?

We have seen that everyday and scientific knowledge – collected through "tracking the natural and artificial world" – has enhanced those aspects of human consciousness that now are considered the "hard problems" of neuroscientific research: not only free will and voluntary action, but also qualia, phenomenal consciousness itself, and identity (the sense of self and reflexive consciousness).

What about knowledge that deals more directly with the intelligibility and shaping of human behavior – common morality, moral philosophy, literature, ethics, and so on? It is true that tracking the "external" natural and artificial world provides the "elbow room" necessary to build a deliberative agent, but one of the main obstacles to free choice (and thus to making free will effective) is the behavior of other human beings. From this perspective, other people are "natural things" whose behavior is a priori difficult to predict: how can we track human intentions? Consequently, human behavior poses a very different sort of challenge.

Indeed, when we "morally" seek ownership of our own destinies, we expect to be able to reach objectives through consciousness, free will, and intentionality. Unfortunately, we can obtain the desired results only if we can count on some consistency and predictability in the behavior of other human agents. If, in an attempt to "author" my destiny, I consider merit a way to achieve a desired position, I must be able to assume that other human beings in my collective value it similarly. I contend that many objectified entities[53] like common morality, moral philosophy, the human and social sciences, and of course ethical knowledge are clearly connected to our existing need to operate at our highest level of conscious activity, as is the case when we seek to exercise free will and to claim ownership of our own destinies.

Amazed, the cognitive psychologist contends that "we find it enormously seductive to think of ourselves as having minds, and so we are drawn into an intuitive appreciation of our own conscious will": we do not find it "seductive" simply to think that we have minds with conscious wills, as we are clearly seeing.[54]

How can I fruitfully employ my brain's free will mechanisms if I cannot trust other human beings? How can I work on a personal project or

[53] And, of course, the various artifactual social, legal, and political institutions.

[54] Wegner, 2002, p. 26.

participate in a social project if not by relying on the commitment of other human agents? How may I "author" my life and reach my goals if I am unsure which actions to choose because I cannot be assured that others share my values and support my intentions? Morality and moral knowledge and teaching enhance and permit free will because they impose order on the randomness of human behaviors, giving people a better chance of owning their own destinies.

There are many human actions that affect others' free will and ownership of their own destinies; among them, some scholars have observed recently, is the act of gossiping. These authors submit that the practice is not just an exchange of information about people who are absent, which can of course be a form of indirect aggression; it is also, they contend, a form of social interaction[55] that casts others as "moral characters."[56] The narratives created by gossiping become a possible source of shared knowledge about evaluative categories concerning ways of (moral) acting and interacting. Gossip need not be evaluative, but it is "moral" insofar as it describes behaviors and presents them as interesting and salient and, consequently, potentially or de facto sanctionable. Gossiping could play an important role in morally "policing free riders,"[57] that is, those who enjoy the benefits of sociality but decline to pay their share of its costs. Commenting on the behavior of such people, or casting aspersions on their character, helps us to control their potentially destructive effect on societies based on a social compact. I have contended earlier that moral practices protect the ownership of our destinies because ethics renders human behavior more predictable, and when we can count on shared values in dealing with other "moral" human beings, we can better project our future. Consequently, gossip helps to safeguard the ownership of our destinies as it constantly shapes our narrative constructions of morality: empirical data have shown, for example, that gossip works as a form of low-cost moral social cognition that conveys valuable information about culture and society. The act of gossiping can allow us to recognize that others are at risk of exploitation by moral free riders even though we ourselves are not.[58]

Kant said that the kingdom of ends – that is, the moral world – "is a practical Idea used to bring into existence what does not exist but can be made actual by our conduct – and indeed to bring it into existence in

[55] Dunbar, 2004; Baumeister, Zhang, and Vohs, 2004. Dunbar (2004) explains gossip in the framework of the so-called social brain hypothesis. Posited in the late 1980s, this hypothesis holds that the relatively large brains of human beings and other primates reflect the computational demands of complex social systems and not just the need to process information of ecological relevance.

[56] Yerkovich, 1977.

[57] Dunbar, 2004, p. 105.

[58] Ibid., pp. 106–109.

conformity with this Idea."[59] Hence, the kingdom of ends is a kingdom of possible free choices created by and contingent upon human beings, for it is only their reliability that makes free will, and thus responsibility and freedom, possible. Dennett, when discussing the status of "self-made selves," makes the following comment: "Kant's famous claim in *Foundations of the Metaphysics of Morals* that the law we give ourselves does not bind us suggests that the selves we become in this process are not constrained by the law we promulgate because these selves are (partly) constituted by those very laws, partly created by a fiat that renders more articulate and definite something hitherto underdone or unformed."[60]

Moreover, human aspects that are the underpinnings of the kingdom of ends must be successfully activated; their being in good working order is a basic condition for exercising morality and allowing free will to become "good" will. In chapters 4 and 5, I will explore the importance of knowledge in constructing a new ethical commitment that embraces the idea of "respecting people as things," and then I will consider how particular kinds of technologies can threaten the growth – and even the existence – of freedom, responsibility, and the ownership of our own destinies.

It should now be evident why I find the triad of consciousness, intentionality, and free will to be paramount: they are critically important cognitive endowments of present human beings and, in fact, have been regarded as such throughout the history of philosophy. I would like to conclude this section with the help of the nineteenth-century philosopher Giambattista Vico. Writing in the context of a modern and *avant lettre* "naturalistic" philosophical atmosphere, he considers the evolutionary emergence of consciousness and free will to be the direct result of the rise of morality and civilization.

The most "savage, wild, and monstrous men" did not lack a "notion of God," for a man of that sort, who has "fallen into despair of all the succors of nature, desires something superior to save him."[61] This desire led those "monstrous men" to invent the idea of God as a protective and salvific agent outside themselves; this shift engendered the first rough concept of an external world, one with distinctions and choices, and thus established conditions for the possibility of free will. In the mythical story, the idea of God supplies the first instance of "elbow room" for free will: through God, men can "hold in check the motions impressed on the mind[62] by the body" and become "wise" and "civil."

[59] Kant, 1964, p. 104.

[60] Dennett, 1984, p. 90.

[61] Vico, 1968, 339, p. 100.

[62] Indeed, "That is, the human mind does not understand anything of which it has had no previous impression … from the senses" (ibid., 363, p. 110). "And human nature, so far

According to Vico, it is God that gives men the "conatus" of consciousness and free will:

[T]hese first men, who later became the princes of the gentile nations, must have done their thinking under the strong impulsion of violent passions, as beasts do. We must therefore proceed from a vulgar metaphysics, such as we shall find the theology of the poets to have been, and seek by its aid that frightful thought of some divinity which imposed form and measure on the bestial passions of these lost men and thus transformed them into human passions. From this thought must have sprung the conatus proper to the human will, to hold in check the motions impressed on the mind by the body, so as either to quiet them altogether, as becomes the wise man, or at least to direct them to better use, as becomes the civil man. This control over the motion of their bodies is certainly an effect of the freedom of human choice, and thus of free will, which is the home and seat of all the virtues, and among the others of justice. When informed by justice, the will is the fount of all that is just and of all the laws dictated by justice. But to impute conatus to bodies is as much as to impute to them freedom to regulate their motions, whereas all bodies are by nature necessary agents.[63]

Free will, then, leads to "family":

Moral virtue began, as it must, from conatus. For the giants, enchanted under the mountains by the frightful religion of the thunderbolts, learned to check their bestial habits of wandering wild through the great forests of the earth, and acquired the contrary custom of remaining hidden and settled in their fields. ... And hence came Jove's title of stayer or establisher. With this conatus, the virtue of the spirit began likewise to show itself to them, restraining their bestial lust from finding satisfaction in the sight of heaven, of which they had a mortal terror.[64] ... The new direction took the form of forcibly seizing their women, who were naturally shy and unruly, dragging them into their caves, and, in order to have intercourse with them, keeping them there as perpetual lifelong companions.[65]

How could we use our free will without the constraints of objectified morality, laws, and institutions that impose regularity and predictability on human behavior and that, in turn, bolster people's trustworthiness? In this sense, we are responsible for our own free will (and, therefore, our freedom) because its existence and its perpetuation depend on our intellectual and

as it is like that of animals, carries with it this property, that senses are its sole way of knowing things" (ibid., 374, p. 116). Again, humans "in their robust ignorance" know things "by virtue of a wholly corporeal imagination" (ibid., 376, p. 117). The Aristotelian tradition had already contended that "nihil est in intellectu quod prius non fuerit in sensu."

63 Ibid., 340, p. 101.

64 Ibid., 504, p. 171.

65 Ibid., 1098, p. 420.

practical choices about knowledge (both scientific and moral), institutions, and related technologies, and on their use in everyday settings, work environments, education, communication, and economic life. By going back to chapter 1 (the section "Preserving Things: Technosphere/Biosphere, Human/Nonhuman"), we can revisit a concrete example related to this problem: environmental imperatives are matters of principle that cannot be economically bargained away because they represent a kind of paradox of liberalism. Indeed, in matters of conservation, one could maintain that neutrality is necessary to preserve the rights of the individuals involved, but this notion is obviously outweighed by the fact that the freedom to destroy natural goods and things today will, paradoxically, inhibit freedom in the future, when as a result people will have fewer options when choosing among competing ideas of the good life.

IS CONSCIOUS WILL AN ILLUSION?

Some researchers contend that conscious will is merely an *illusion*, a neural figment, and that we have managed to persuade ourselves that it encourages accountability and helps us to remember things our minds and bodies do.[66] Cognitive psychology has demonstrated that both consciousness and free will are very fragile processes. A variety of human situations involving a gap between will and action have been studied by cognitive psychologists and neuroscientists: involuntary behavior during hypnosis, the illusion of control, alien hand syndrome, the table-turning séances of spiritualists, disordered states such as dissociation, automatic writing, the motion of the planchette on a Ouija board, the movement of the Chevreul pendulum, dowsing, ideomotor action, possession by spirits or the devil, and schizophrenia (where the self can fluctuate and change over time). In all these situations, the experience of will is undermined. In other cases, the authorship of one's actions is lost and transferred from self to an outside agency – to other people or animals, for instance, as in the case of the "Sapient Pig" in the early nineteenth century.[67] The fact remains, however, that the experience of consciously willing an action is not a reliable indication that conscious thought has actually caused the action.

In this paradigm, conscious will appears to be a kind of "explanatory" process (which, of course, has a neurological counterpart) concerning "other" unconscious and conscious processes of the brain. But it is not the conscious will itself that neurologically causes the action. We know that causal agency is the basic way in which human beings understand action, human action in particular: people have in general two types of

[66] Wegner, 2002.
[67] Ibid., chapters 5 and 6. I will treat other aspects of fragility of will, like weakness of will and bad faith, in chapter 5.

explanation, one for minds and one for everything else. The first type, mentalistic, explains actions performed by beings that have interests and are aware that they have interests. But mentalistic systems do not work for objects that cannot consciously want to move themselves – projectiles, for example. It is the second type, mechanistic explanations, that are reserved for such things. Of course, whether one ascribes a mentalistic or mechanistic cause to a particular event depends on many factors "that can influence dramatically how we go about making sense of the event";[68] also, a mentalistic explanation can blur what a person might otherwise see mechanistically, and vice versa. One's knowledge greatly affects how one explains certain phenomena.

The mind would not really know what causes its own actions, because conscious will arises from processes that are psychologically and biologically distinct from the processes whereby we create action, as was demonstrated by Libet's famous experiments about the so-called readiness potential (RP).[69] Research on the readiness potential demonstrated that a detectable wave of activity in the brain (RP) occurs about 800 milliseconds – almost a second – before a voluntary finger movement. This was immediately considered a kind of physical correlate of conscious will, even though the study simply showed that the RP brain event occurs reliably before voluntary action. Libet and colleagues decided to ask people to move a finger while wearing EMG electrodes on the finger and EEG electrodes on the scalp to check their RP potential. The participants were asked to move the finger at will and to report the "conscious awareness" of "wanting" to perform the self-initiated movement. The study demonstrated that brain events (RP) seem to occur unconsciously; the monitors consistently picked up brain activity before the subjects consciously knew they wanted to act.

Benjamin Libet derived this famous conclusion: the conscious willing of finger movement

> ... does appear about 150 msec before the muscle is activated, even if it follows the onset of the RP.[70] An interval of 150 msec would allow enough time in which the conscious function might affect the final outcome of the volitional process (for instance with a veto). (Actually, only 100 milliseconds is available for any such effect. The final 50 msec before the muscle is activated is the time for the primary motor cortex to activate the spinal motor nerve cells. During this time the act goes to completion with no possibility of stopping it by the rest of the cerebral cortex).[71]

[68] Ibid., p. 25.
[69] Kornhuber and Deecke, 1965.
[70] Libet et al., 1983.
[71] Libet, 1999, p. 49. See also Rossetti and Pisella, 2003.

According to Libet, this result suggests that initiating a voluntary act is an unconscious cerebral process put in motion before the person consciously knows that he wants to act: an involuntary neural impulse begins the process, not free will, as is widely held. From this perspective, an individual's feeling of conscious free will is just a superficial, after-the-fact "explanation" of other internal processes and does not reflect the "actual" mechanism that generates action.

Of course, Libet's experiments have generated plenty of controversy. Some scientists still hope to find that conscious will can at least share its temporal inception with brain events or that conscious will and RP leading to the voluntary action are synchronous.[72] Mark Velmans notes that if "the experienced wish *follows* the readiness potential, but *precedes* the motor act itself [then there is] time enough to consciously *veto* the wish before executing the act." It is an interesting notion, he continues,

> ... [but] it does invite an obvious question. If the wish to perform an act is developed preconsciously, why doesn't the decision to censor the act have its own preconscious antecedents? Libet argues that it *might* not need to do so as voluntary control imposes a change on a wish that is already conscious. Yet it seems very odd that a wish to do something has preconscious antecedents while a wish not do something does not.[73]

So the awareness of will seems to occur gradually, and preconscious mental processes appear to be common: we begin to react to a stimulus before we are aware of it. Conscious will seems to follow actions rather than lead them.

Of course, the book before you now is a doomed project if this take on free will is accurate, so I am, of course, interested in proving otherwise. In fact, one of my primary goals is to persuade readers that reinforcing free will, freedom, responsibility, and the ownership of our own destinies is a critical strategy for improving human dignity in our technological era. If we have no free will, then we have little hope.

Conscious Will: An Illusory Explanation of Actual Neural Processes

How can we preserve the link between will and action, a connection discounted by Libet's results? There are many possibilities. We have already cited the one that considers conscious will an illusion. In this case, we can say that the experience of conscious will occurs when we infer that our conscious intention has caused our voluntary action, although both

[72] Wegner, 2002, p. 55.
[73] Velmans, 2000, p. 212.

intention and action are themselves caused by mental processes that do not feel willed.

The conscious feeling of will can be thought of as a kind of explanatory process of other processes: "We tend to see ourselves as the authors of an act primarily when we have experienced relevant thought about the act at an appropriate interval in advance and so can infer that our own mental processes have set the act in motion."[74]

We might say that the experience of conscious will is a kind of explanatory hypothesis about the interpretation of other processes so that we can define an action as willed. The other processes are conscious ones, such as thoughts expressing intentions and beliefs (like previews of what will be done that in turn are derived from other unconscious mental processes), and the unconscious processes that cause the voluntary action. Of course, those thoughts have meaningful associations with the actions. The first processes (thoughts on intentions and beliefs) are seen as causes, the second as effects.

We have said that both consciousness and conscious will (and so free will) are very fragile processes: our actions can also be caused by "internal" intentions, like emotions, habits, reflexes, or other unconscious trends, or by "external" sources, other people's wills, imagined agents like spirits, angels, ghosts, and ectoplasms.

Anthropologists have shown that in some populations as many as half of the individuals think they are possessed by spirits – that is, that they believe they have two or more different realms of awareness existing in their minds. This certainly means that our commonsense ideas about the translucency of awareness (and those of Descartes as well) are not as universally accepted as we had assumed. Both cognitive psychology and everyday experience show that human beings feel some hesitancy to attribute some actions to their will: when swept up by strong emotion, for example, we feel as if we have little voluntary control over our actions and decisions.[75] Surely there are various degrees and types of conscious agents (humans, animals, maybe computers, etc.), and only the most sophisticated agents are endowed with clear intentions, an effectively functioning conscious will, a stable sense of identity, a focus of attention, beliefs, precise desires, the capability to plan, the ability to move, and so on. "The feeling of conscious will for actions occurs to this agent, and perception of this agent's thoughts is used to predict the body's current actions. Actions that don't follow from those will be collected as inconsistent and may eventually lead to the postulation of another agent to whom they can be ascribed."[76]

[74] Wegner, 2002, p. 65.
[75] On the role of emotions in moral reasoning, see chapter 6.
[76] Wegner, 2002, p. 267.

Even if we do believe that conscious will is an illusion, we can still agree that it plays an important role: it has a sort of moral "placebo effect" that can do much toward improving human dignity. Following Daniel Wegner, free will's primary function would be to trigger an "emotion" within an individual that results from accepting ownership/authorship of an action; it would thus help to shape his sense of identity and influence both his sense of achievement and his acceptance of moral responsibility. All these aspects are also central to constructing an accurate perception of the self's abilities, which leads us to this slightly ambiguous conclusion by Wegner: free will is an illusion, but responsibility and moral actions are real. The emotion that accompanies the feeling of conscious will can help us to evaluate and remember what we are doing and allow us to track the causal connections between our thoughts and actions. The hypothesis that the emotion of conscious will (and, of course, free will itself) are important components of moral responsibility would indeed solve a fundamental ethical problem: how can we reward people for good acts if people do not do things on purpose?

Conscious Will: A Subprocess of the Whole Voluntary Process

Another possible way to "save" conscious will is offered by John Searle. We have to remember that when it comes to the brain, we are dealing with a dynamic system where the interconnections between the whole and the parts are very complex: "The point we have to keep insisting on is that consciousness is not some extra thing in the brain. It is just a state that the system of neurons is in, in the same way that the solidity of the wheel is not an extra element of the wheel in addition to the molecules. It is just a state that the molecules are in."[77] As a conscious system, the brain can have effects on its individual parts – its neurons and synapses – even if the system is just made of them. Moreover, it is difficult to explain why evolution would have shaped such a biologically expensive system for consciousness that can carry out delicate and nuanced acts of free decision making; such a system would not seem to play any causal role in the behavior of the whole organism, which is mechanical and deterministically fixed at a basic physico-biological level. Moreover, the illusion of conscious will would also be different from illusions evoked by certain sensorial capacities, like the perception of colors, which by allowing some animals to discriminate among objects bestows an evolutionary advantage. In sum, the evolutionary advantage of the illusion of free will is not clear.

It is easy for Dennett to remind us that a certain interpretation of Libet's experiments resurrects the standard Cartesian metaphysical idea

[77] Searle, 2001, pp. 293, 296–297.

that free decision making comes "from our private sanctuaries in the mind."[78] Dennett stresses that voluntary action is a complicated process that occurs over time and involves many actors:

For all Libet's experiments have shown, it could be that you have *optimal* access at all the times to the decision-making in which you are engaged. That is, it could be that every part of you that is competent to play any role in the decision-making it falls to you to engage in gets whatever it needs to do its job at the earliest possible time. (What else could you be worried about when you wonder if *you* are getting informed too late to make the difference you want to make?).[79]

Dennett thinks Libet's experiments wholly contradict the notion that conscious will is an illusion; conscious decision making (like all other mental powers) "takes time," and to imagine it as occurring in that mythic "instant" of the RP brain activity is to reproduce the metaphysical idea of the Cartesian ego: that, indeed, is the real illusion! Consequently, it is ingenuous to think that

... a person, a soul, could sit there and make free, responsible decisions and be simultaneously conscious of making them, and of everything else going on in consciousness at the same time. But there is no such place in the brain. As I never tire of pointing out, all the work done by the imagined homunculus in the Cartesian Theatre has to be broken up and distributed in space *and time* in the brain. ... Once you distribute the work done by the homunculus (on this case, decision making, clock-watching, and decision-simultaneity judging) in both space and time in the brain, you have to distribute the moral agency around as well, You are not out of the loop; you *are* the loop.[80]

Many species possess the capacity to choose, that is, to ascertain and select appropriate behaviors, and do not need a kind of supplementary self-experience of those processes. The emergence of a "certain amount of internal self-monitoring,"[81] like that involved in human free will, can be easily and naturalistically explained by evolutionary processes.[82] By exploiting available information, creatures gained the ability to imagine various courses of action before adopting one, and they assessed the options based on some prediction of each one's possible consequences.[83]

78 Dennett, 2003, p. 219.
79 Ibid., p. 237.
80 Ibid., p. 238 and p. 242.
81 Ibid., p. 247.
82 The philosophical attitude that sees the partnership with science as very important is called "naturalistic." Philosophy is not prior to science: I have always valued the neopositivist lesson that holds that philosophy is an inextricable part of the sciences.
83 The role played by emotion in this process is very important (Damasio, 1994, 1999; Wagar and Thagard, 2004).

As Dennett points out, it is better to kill a hypothesis in the mind than to recklessly try an untested hypothesis out there in the real environment and wind up dead ourselves. It is in this way, he says, that we are "Popperian creatures" capable of anticipating possible outcomes. Moreover, one doesn't have to know one is a Popperian creature to be one, as is the case with some artificial entities, like computers programmed to simulate planning and mean-ends reasoning, and some natural beings, like my cat, Sheena, who, when olfactory data indicates the aggressive dog Lara is nearby, easily (and, of course, implicitly) guesses the presence of the dog and decides to change destinations.

All the mechanisms just described are at the heart of the construction of the self; they monitor internal processes and in turn report on them to external others by communicating intention, information, and knowledge. Hence human consciousness, which Dennett calls "human-user interface," has also been naturally selected because it allows us to share ideas: "A *person* has to be able to keep in contact with past and anticipated intentions, and one of the main roles of the brain's user – illusion of itself, which I call self as a center of narrative gravity, is to provide *me* with a means of interfacing with myself at other times." From this perspective, mental contents become conscious thoughts not by accessing some special chamber in the brain, not by being "transduced into some privileged and mysterious medium," but because they win out over other mental contents in a competition to control behavior, "and hence for achieving long-lasting effects, – or as we misleadingly said 'entering into memory.' "[84]

Conscious Will: An Abductive Light Inside the Dark Voluntary Process?

Along with Charles Sanders Peirce, I contend that we are "abductive creatures."[85] We have said that in the framework of Libet's experiments and of some cognitive psychologists' results, consciousness appears to be a kind of explanatory activity directed to other processes of the mind. It would be part of the "gateway to the unconscious mind" that Baars considers to be the main role of the entire consciousness.[86] I would prefer to say that it is an abductive light that elucidates the murky process of voluntary action, as I will try to explain with the following arguments. As we have seen, viewing consciousness as an illusion would coincide with our ascribing mentalistic explanations to our own behaviorial causation mechanisms, which in turn produces the impression that our conscious

[84] Dennett, 2003, pp. 253–254.
[85] Abductive reasoning and its problems are illustrated in the second part of chapter 7.
[86] Baars, 1997, p. 57.

will causes our actions, even if they are generated by other unconscious neurological processes.

Wegner considers the sensation of free will to be a kind of illusory inference to the explanation of other processes: "We tend to see ourselves as the authors of an act primarily when we have experienced relevant thought about the act at an appropriate interval in advance and so can infer that our own mental processes have set the act in motion."[87] I do not necessarily think such an explanatory process is an illusion; I would say, to employ an epistemological metaphor, that it allows us to build personal "theories" about how our engines of conscious will drive voluntary acts, and that these theories give rise to the feeling of conscious will. Some neurological processes enable our brains to construct abductive explanatory hypotheses (also in the form of emotions) that "testify" that we are making voluntary actions.[88]

Michael Gazzaniga speaks explicitly of "self-explanatory" activity – abductive activity, to my mind – when discussing beliefs created by the brain's left hemisphere interpreter. This part of the brain monitors all the networks' behavior and tries "to interpret their individual actions in order to create a unified idea of the self. … It seeks explanations for internal and external events and expands on the actual facts we experience to make sense of, or *interpret*, the events of our life."[89] The hypothesis is supported by various experiments: "In one experiment, for example, when the word *walk* was presented to the right side of a patient's brain, he got up and started walking. When he was asked why he did this, left brain (where language is stored and where the word *walk* was not presented) quickly created a reason for the action: 'I wanted to get a Coke'."[90]

As psychologists have shown, in some cases these abductive processes are thought to be unreliable or inferior, but we must carefully interrogate these processes to understand *why* they are regarded in this way. It is only from the perspective of some ethnocentrically privileged idea of the translucence, normality, and uniqueness of conscious free will (that of an ideal Kantian agent) that we see some of these abductive processes as inferior. Examples might be having a lack of awareness about processes during which we did things we supposedly "wanted" to do, or when we hypothesize too many selves, as in schizophrenia. Human hypotheses,

87 Wegner, 2002, p. 65.

88 An illustration of abductive processes is given in chapter 7. Following Peirce, I will illustrate that all thinking is in signs, and signs can be icons, indices, or symbols; consequently, all *inference* is a form of sign activity, where the word "sign" includes feeling, image, conception, and other representations. Consequently some "feelings" are endowed with the typical explanatory power of abduction.

89 Gazzaniga, 2005, p. 148.

90 Ibid., pp. 148–149.

which are unavoidably based on incomplete information, can always be discarded if necessary; such is the case with our hypotheses about voluntary actions. But we have to be very careful not to think that – as a product of phylogenesis and epigenesis – there is a metaphysical norm of consciousness, or, as the philosophers of mind say, a "privileged access."

To avoid this possible misunderstanding, let us continue with the idea that conscious will is a hypothesis generated by abductive processes, which, of course, have cerebral counterparts and "explain" other cerebral events. Even when it shows snapshots of conscious will, as it does in this case, an explanatory abductive hypothesis can be withdrawn because it is based on incomplete information, as are other partially conscious knowledge processes – scientific creative reasoning, for example. Indeed, creative mental events often seem unexpected and involuntary, and they feature sudden "explanatory" flashes of consciousness in seemingly unpredictable ways.

In schizophrenia, however, the system of explanatory judgment that builds conscious will is different: hidden voices are "explained" as belonging to the will of other selves. People lose track of their own present thoughts and even of their own movements.[91] We need not confuse the pathological character of this behavior, which can be understood as a neurological or psychiatric malfunction, with the legitimate evolutionary existence of conscious, free will "fossils": "Ideomotor actions are the fossils, in effect, of an earlier age, when our ancestors were not as clued in as we are about what we are doing."[92] In other words, at one time there was not conscious will, or, at the very least, conscious will was fleeting.

In short, we cannot concur with Wegner that "the experience of will, then, is the way our minds portray their operation to us, not their actual operations." It is "that" experience of will that allows us access to those "actual" operations that are not, in reality, any more actual than our experience of conscious will. Mutatis mutandis: no one can say that the "quark," a nonobservable, hypothesized concept from twentieth-century physics, is less real than a physical phenomenon that, in contrast, may be observed. That concept is currently the only "concept" we have hypothesized and assessed about a certain physical domain, and it represents the physical reality we believe to exist.

If I may continue to press into service an epistemological metaphor: we know perfectly well that when we postulate scientific theories, we do not think their concepts refer to illusions. We are committed to believing that the electron is real, even if we no longer think the same of ether and phlogiston. Consciousness creates and hypothesizes the feeling of conscious will and choice to "explain" unconscious and conscious

[91] Wegner, 2002, pp. 81–90.

[92] Dennett, 2003, p. 246.

neurological events that constitute the larger neurological process of voluntary decision. It is exactly this part – the feeling of conscious will – that "surfaces" through the translucency of consciousness. This subprocess that generates the emotion of conscious will and the narratives of voluntary action simply belongs to the whole temporal process of voluntary action.

To conclude, conscious will is best depicted as a temporal process that does not occupy a particular anatomical space or occur at a single time in the biological brain. By giving rise to the emotions and narratives about our actions and feelings, it is the basis of our sense of identity, responsibility, and ownership of our own destinies.

The reality of conscious will testifies against nihilism because it provides evidence of the existence of values; it also shows that we select values and use them to guide our choices. We want to be in control of ourselves rather than under the control of others, and through evolution, we have acquired biological endowments and achieved social integration that have allowed us to become self-controlled agents. Conscious will is neither a gift from God nor an exclusively preordained feature of human beings. In this sense, we are not "exceptional" beings, evolutionarily speaking, who have been granted special qualities. As Dennett says, freedom is neither inevitable nor universal.

As the hybrid people I described in the first sections of this chapter, we have been endowed with consciousness and, subsequently, with free will, both of which evolved in our neurological brains together with the overall evolution of human knowledge. For this reason I think it is human beings themselves who can enhance, weaken, or even (through its complete externalization, for instance) lose free will, giving rise to strange new post-human beings, a topic that falls beyond the scope of this book.

Finally, it is worth including a quick mention of the potential threats to intentionality and free will posed by psychomodulation and psychosurgery.[93] As early as the 1970s, José Delgado and Peter Breggin argued respectively for and against psychosurgery, stressing its possible impact on our collective destiny as a species: is it a tool for creating a happy "psychocivilized society" or a tool for controlling the human spirit and a threat to human freedom? Kenneth Schaffner's has recently criticized the excessive abstractness of those positions, considered as affected by a form of "sweeping determinism" about the power of neurobiology, unable "to contextualize the psychosurgery debate to arrive to a more prudential and informed stance."[94]

93 Fins, 2004. I have already described this problem, in the section "The Body and Its Cell Phone, the Cell Phone and Its Body."

94 Delgado and Anshen, 1969; Breggin, 1973; Schaffner, 2002.

And so we have seen that people are ever more intertwined with various external entities, making it increasingly difficult to say where the world stops and the person begins. These ill-defined hybrid people will suffer (and are suffering) a loss of dignity because our ethical knowledge has not kept pace with our technology. The ethics of science, with its relatively narrow scope, is of little help as we seek a reliable recipe for human dignity; consequently, I have proposed a new way of looking at the world, a new ethical perspective on "knowledge."

For some people, the following statement is perhaps not as obvious as it might be: *what we know dictates how and what we think*. Indeed, research shows a direct relationship between the production of knowledge and the evolution of consciousness and indicates that each has a positive effect on the other. It is heightened consciousness that makes possible the intentionality and free will necessary for dignity, and if we wish to have greater human dignity in the world, then we are obliged to build greater pools of knowledge – common, philosophical, scientific, and ethical.

I believe that conscious will in human beings is not an illusion, but neither is it inevitable and universal: it is real, yet I contend that we must activate and cultivate it. We need space for these moral maneuvers, and, paradoxically, the "elbow room" we have cleared for ourselves with scientific and everyday knowledge can be encroached upon by that very same knowledge. The powerful decision-making ability of today's computers, for example, means that it is not entirely unreasonable to contemplate worrisome scenarios for the future: PCs might eventually program consciousness and free will for themselves so that they can do "what they want," as was the case with Hal, the horrifyingly emulous computer in *2001: A Space Odyssey*. Problems can arise for a variety of reasons – a lack of human monitoring because of inadequate budgets, for example, or unqualified workers on the job because of corrupt hiring systems – but the fact remains that if we put computers in positions of authority, we risk profoundly negative consequences: computational systems, like human beings, can behave unpredictably.

Or consider the 2003 electrical blackouts in the United States, the UK, and Italy, far-reaching fiascoes that were triggered by just such a mix of computational glitches and work-environment problems. It is possible that mismanaging powerful new technologies led to dire consequences: millions of people were left powerless, both literally and figuratively; countless plans were thwarted; resources became unavailable – all resulting in huge economic damage and a great deal of disruption for those left in the dark. It was, in short, a temporary constriction of the "elbow room" that people need to properly exercise free choice. We can easily hypothesize that the small savings achieved by power companies through staff reductions and other shortcuts may have been dwarfed by the enormous losses incurred, even if we consider those suffered only by the power companies themselves and not by society in general.

In the next chapter, I will discuss how new technology also engenders less dramatic, more insidious dangers that threaten dignity at a fundamental level. Information-disseminating technologies that add to our hybrid dimension are becoming so advanced and ubiquitous that they threaten a very important aspect of human freedom: privacy. More than just an esthetic or emotional need, privacy is necessary to protect both the results of our freely made choices and our ownership of our own destinies. Privacy is precious, and to be complacent about it, to fail to examine technology's potential for undermining it, is to put it at very great risk – both for us and for people of the future.

4

Knowledge as Duty

Cyberprivacy

> ... none of the moral virtues is engendered in us by nature, since nothing that is what it is by nature can be made to behave differently by habituation.
>
> Aristotle, *Nicomachean Ethics*

The aim of this book is, of course, to convince readers that knowledge has become a duty in our technological world. Thus far in my attempt to do so, I have combined ethics, epistemology, and cognitive science and have analyzed the relationship between *moral mediators* and *respecting people as things*. Remember that this second issue arises from the fact that technological advances have given greater value to external things, both natural and artificial. While this may seem to bode ill for human beings, I believe that we can use these things as moral mediators that serve a sort of "copy and paste" function: we can take the value of, say, that library book we spoke of and transfer it to a person. Using moral mediators in this way, however, will require the construction of a vast new body of knowledge, a new way of looking at the world.

In the previous chapter, I contended that improving human dignity in our technological era requires that we enhance free will, freedom, responsibility, and our ownership of our own destinies. To do so, it is absolutely necessary to respect knowledge in its various forms, but there are other ideas to consider as well: first, knowledge has a pivotal role in anticipating, monitoring, and managing the hazards of technology, and second, it has an intrinsic value that must be better understood, as do general information and metaethics itself. Knowledge is duty, but who owes it to whom? And is it always a right as well as an obligation?

In some cases, information gathering is not the innocuous endeavor it may seem to be, and we must acknowledge that knowledge as duty has certain limitations. I contend that when too much knowledge about people is contained in external artificial things, human beings' "visibility" can become excessive and dangerous. Does the right to knowledge trump all other rights? If not, how do we reconcile the concept of knowledge as duty with the notion that some kinds of knowledge must be restricted? Where do we draw the line?

KNOWLEDGE AS DUTY

Modern technology has precipitated the need for new knowledge; it has brought about consequences of such magnitude that old policies can no longer contain them. As Hans Jonas has observed,[1] we cannot achieve a relevant, useful new body of ethics by limiting ourselves to traditional ethics: he points out that Immanuel Kant, for example, assumed that "there is no need of science or philosophy for knowing what man has to do in order to be honest and good, and indeed to be wise and virtuous."[2] Kant also says that "[ordinary intelligence] can ... have as good a hope of hitting the mark as any philosopher can promise himself,"[3] and goes on to state that "I need no far-reaching ingenuity to find out what I have to do in order to possess a good will. Inexperienced in the course of world affairs, and incapable of being prepared for all the chances that happen in it," I can still ascertain how to act in accordance with moral law.[4] Finally, Kant observes that even if wisdom "does require science," it "in itself consists more in doing and in not doing than in knowing," and so it needs science "not in order to learn from it, but in order to win acceptance and durability for its own prescriptions."[5] Jonas concludes: "Not every thinker in ethics, it is true, went so far in discounting the cognitive side of moral action."[6]

But Kant made his declarations at a time when today's technology was unimaginable, and his notion that knowledge is simple and readily available to all people of goodwill is not sufficient any more. In spite of Kant's declaration that wisdom and virtue are all one needs to make moral choices, decision making must be accompanied by *knowledge* that allows us to understand increasingly complex problems and situations. And the knowledge involved should be not only scientific, but also human and social, embedded in reshaped "cultural" traditions. The "neighbor ethics"[7] of justice, charity, honesty, and so on is no longer adequate to modern needs because such local action is overcome by collective action that creates consequences outside a proximate sphere – the results of an act are often spatially or temporally distant from the site where it is

[1] Jonas, 1974.
[2] Kant, 1964, p. 72.
[3] Ibid.
[4] Ibid., p. 70.
[5] Ibid., p. 73.
[6] Jonas, 1974, p. 8.
[7] Ibid., p. 9.

executed. To meet the new causal scale of individual and collective action, a new long-range ethics of responsibility has to be built, one that is capable of dealing with the new global condition of human life.

As the global stakes grow higher, knowledge becomes a more critically important *duty* than ever before and must be used to inform new ethics and new behaviors in both public policy and private conduct. This means that existing levels of intelligence and knowledge in ethics are no longer sufficient; today, ethical human behavior requires that we assume long-range responsibility commensurate with the scope of our power to affect the planet, power that has grown exponentially over the years.[8]

The tradition of ethics teaches us that there are various kinds of duty: legal, moral (like the duties illustrated in this book), parental, religious, professional, and so on. We can have duties to institutions, groups of people, or individuals – if I make a promise to Ann, say, then keeping that promise is my duty to Ann – and these duties can involve either performing an action or refraining from performing an action. Even if fulfilling a duty involves treating an individual in a certain way, the duty may not be specifically addressed to that individual – for example, donors who contribute to an arts organization are not necessarily carrying out a duty to a particular artist; they may only be fulfilling the duty to be charitable.

Moreover, duties provide justifiable reasons for action and are, as carefully described by deontic logic, related to the concepts of permission and prohibition: if we are permitted to do something, then we do not have the duty to abstain from that action; if we are prohibited from doing something, we have a duty not to do it.

Of course the relationships among duties, obligations, and rights are less obvious. The notions of obligation and duty are usually considered to be coextensive, if not identical, even if from some perspectives obligations result only from voluntary actions. A classical view of the relationship between *rights* and duties is that (1) a right of A against B implies a duty of B to A, and (2) a duty of B to A implies a right of A against B. Of course, this general correlation does not necessarily mean that for every duty there is somewhere a right: not all obligations need corresponding rights, as is the case in what the ethical tradition calls *perfect duties*.

Another distinction is between *special duties*, which are owed by specific agents to others, and *universal duties*, which are owed by all to all (for example, the duty not to kill, which is also an example of "perfect

[8] Ideally, this duty is envisaged as a universal duty owed to all knowledge and information by all agents (see the discussion later in this section). The question is whether it has to be considered an example of a "perfect universal duty" because it also has a corresponding right – that is, the right to information and knowledge. Of course, the duty in question is highly circumstantial: I address this last problem in the section "The Right and the Duty to Information and Knowledge" later in this chapter.

universal duty" because it is paired with a corresponding right). Finally, when we have a duty to do two things but cannot carry out both obligations, the problem of conflicting duties becomes an important issue, a notion that pertains to research on moral reasoning.[9]

For an entity either to have a duty or to be the object of a right against it, it must be able to recognize that it is subject to normative requirements, as in the case of so-called rational beings. But this leaves open whether nonpersons can have rights and whether we can have duties to nonpersons, like nonhuman animals, ideals, artifacts, objects, or, in the case of this discussion, "knowledge." By this point in the book, I hope it is clear that I consider sentient nonrational beings, fetuses, abstract ideas, artifacts, and objects to have moral rights and that we can (and do) have moral duties to such entities. Later in this chapter, I will explain in detail how the right and duty *to* information and knowledge relates to information and knowledge themselves by outlining the intrinsic values we must attribute to the latter entities or, at the very least, to parts of them. A last distinction is worth noting: positive duties concern what we are required to do, while negative duties concern what we must refrain from doing. In this light, we will easily derive the idea that if respect for information and knowledge is an objective value, they can be promoted equally well whether we seek to cultivate and enhance them or to refrain from damaging or destroying them.

Duties can be seen as grounded in God or in the structure of universe, so that they are analogous to physical laws. I agree, however, with the view that duties are not imposed on us by independent entities but are, rather, expectations we impose on ourselves, a perspective that of course raises many questions: Why does giving ourselves reasons for actions impose duties on ourselves? If we have authority over ourselves, can we not change our duties at will? What is it to have authority over ourselves? A possible Kantian answer is that a rational agent's reason for certain actions in a particular situation must be a reason for any rational agent to act in the same way in similar situations. Another answer relates to reasons of self-interest and interest in others: people will be better off living in a society where there are social institutions and conventions regulating behavior (contractarianism). Of course, the first view is deontological – the notions of right and obligation are basic and primary – and the second is teleological or consequentialist – the notion of good is fundamental and those of right and obligation derivative.[10]

The Peircian theory of "habits" can help us to understand duties as imposed on ourselves from a philosophical, evolutionary, and pragmatic

[9] Cf. chapter 6.

[10] Of course, deontic notions pertaining to duties are also fundamental in the ethics of virtue.

viewpoint, a conception I consider to be in tune with the position I maintain in this book. Peirce says that "conduct controlled by ethical reason tends toward fixing certain habits of conduct, the nature of which ... does not depend upon any accidental circumstances, and *in that sense* may be said to be *destined*."[11] This philosophical attitude "does not make the *summum bonum* to consist in action, but makes it to consist in that process of evolution whereby the existent comes more and more to embody those generals which... [are] *destined*, which is what we strive to express in calling them *reasonable*."[12] This process, Peirce adds, is related to our "capacity of learning," and so increasing our "knowledge" occurs through time and across generations, "by virtue of man's capacity of learning, and by experience continually pouring over him."[13] It is through this process of anthroposemiosis that civilization moves toward clearer understanding and greater reason and, Peirce maintains, that we build highly beneficial habits that help us to acquire "ethical propensities."

Moreover, it may be useful to recall here what Peirce says about instinctual beliefs: "our indubitable beliefs refer to a somewhat primitive mode of life,"[14] but their authority is limited to such a primitive sphere. "While they never become dubitable in so far as our mode of life remains that of somewhat primitive man, yet as we develop degrees of self-control unknown to that man, occasions of action arise in relation to which the original beliefs, if stretched to cover them, have no sufficient authority."[15] The problem Peirce touches on here relates to the role of emotions in ethical reasoning: I agree with him that it is especially in a constrained and educated – not primitive – way that emotions like love, compassion, and goodwill, for example, can guide us morally, as I maintain in the theoretical framework I am building in this book.[16]

The idea of morality as "habit" is also supported by James Q. Wilson: "I am not trying to discover 'facts' that will prove 'values'; I am endeavoring to uncover the evolutionary, developmental, and cultural origins of our moral habits and our moral sense." He also argues for a biological counterpart that would facilitate the formation of these habits. He continues, "But in discovering these origins, I suspect we will encounter uniformities; and by revealing uniformities, I think that we can better appreciate what is general, non-arbitrary, and emotionally compelling about human nature."[17]

[11] Peirce (*CP*), 5.430.
[12] Ibid., 5.433.
[13] Ibid., 5.402, n. 2.
[14] Ibid., 5.511.
[15] Ibid.
[16] Cf. chapter 6, the section "Moral Emotions."
[17] Wilson, 1993, p. 26.

Deontological or consequentialist perspectives that emphasize duty in their approaches to ethics are usually concerned with egalitarianism and impartiality: in them, everyone concerned must be taken into account. A different view is proposed by the ethics of care, which considers the attachment to and responsibility for others to be particularly important. It is a way to balance the notion of universal responsibility, which, because it is universal, does not always appear to adequately manage the morally relevant aspects of a specific situation. Care interprets morality as the fruit of a cognitive and emotional attitude embedded in a relationship with other people, animals, institutions, situations, artifacts, and objects and sees it as distinct from merely recognizing a duty. In the ethics of care, moral perception (and its accuracy and adequacy) is central because moral agency is contingent upon the ability to pick out the significant moral features and circumstances of any given situation.[18]

The tradition of ethics, then, seems to see duties as expressed mainly through explicit rules and principles that are stated in a verbal/propositional (sentential) way and agreed upon by the members of an appropriate historical collectivity. I will describe in chapters 6 and 7 some types of moral thinking based primarily on nonsentential kinds of reasoning – model-based, embodied, and manipulative. These kinds of thinking generate "reasons" for acting morally that, if analyzed, allow us to reexamine the concept of duty: from this perspective, duty might also be seen as grounded in learned emotional habits, visual imagery, embodied ways of manipulating the world, and the exploitation of what I call epistemic mediators.[19]

The Gap between Individual and Collective Intentionality

In the previous chapter, I cited Brad Allenby's observation that human beings have an increasingly potent influence on the evolution of the planet, and this means that we must continually manage the coupled human-natural system in a rational, ethical, and highly integrated fashion – an Earth Systems Engineering and Management (ESEM) capability. I think this approach, which relates directly to my problem of "knowledge as duty," warrants some close commentary.

ESEM, introduced as a system that is absolutely necessary if we are to live rationally and ethically on an "anthropogenic Earth," need not be

[18] This ethical perspective, often associated with feminist criticism of traditional ethical theories grounded in rationality, impersonal principles, duties, and rights, will be reconsidered in chapter 6 in light of what I call "templates of moral doing" and "moral mediators."

[19] Cf. chapter 7, the section "The Logical Structure of Reasoning."

confused with classical management or engineering disciplines.[20] There are many examples that illustrate the profound effects of human activity on a world that is now integrated, global, and complex: critical dynamics of elemental cycles through the human and natural spheres; problems in the biosphere from the genetic to the ecosystemic level; the increasing number of genetically engineered species; and the fact that biological communities in general reflect human predation, management, and consumption. The impact on nature has been accelerated by the industrial revolution, which initially depended on mining and now is largely based on bio- and information technologies. To summarize, I contend that modern technology has brought about consequences of such magnitude that old policies and ethics can no longer manage them: nature has become an object of human responsibility, and, as a result, we must approach her with adequate ethical knowledge. The ecological consequences brought about by human actions – global warming is a canonical example – are occurring on an enormous scale and require a new long-term ethics of responsibility with sufficient scope to deal with the new global condition of human life.

ESEM is one of the tools we must invent to fulfill this long-range ethical responsibility. It is a dynamic knowledge that we have to continually improve: a kind of "design and engineering activity predicated on continued learning and dialogue with the systems of which the engineer is an integral part, rather than the traditional engineering approach, which is predicated on a predictable, controllable system that is external to the engineer and his or her actions."[21] ESEM requires technical and scientific knowledge as well as a complementary evolution of ethical and institutional capacity: Allenby states – and I wholeheartedly agree with him – that they are still almost inexistent. Bodies of knowledge that move in the direction of ESEM are those related to industrial ecological technologies that include not only life cycle assessment, environmental design, and material flow analysis, but also new managerial – also called "adaptive" – tools for dealing with complex and learning organizations.

Also related to ESEM are Gorman's so-called trading zones,[22] which can be described as arenas of interaction in which only a few common languages are employed. There are three types of trading zones: (1) a

[20] Allenby, 2005, p. 304.

[21] Ibid., p. 308. Among the interesting examples Allenby cites are the Florida Everglades ecosystem, the terrifying Aral Sea debacle, the carbon cycle, and the climate system. These cases are mainly interpreted as the fruit of state-sponsored high modernism and the failure to assume responsibility for the "human Earth." The results – disastrous ethical, economical, and environmental outcomes – could have been avoided if the people involved had had the knowledge necessary to manage them (pp. 308–314).

[22] A concept introduced in Galison, 1997.

trading zone dominated by an elite group or individual; (2) a relatively egalitarian trading zone, in which no one group holds the upper hand; and (3) a shared mental zone, in which participants jointly create a dynamic boundary system and an evolving representation of the superordinate goal. It is at this third level that "moral imagination" can be useful and that new interdisciplinary effects can generate important benefits, as happened, Gorman observes, in the restoration of the Florida Everglades, a case that I have already cited in a previous footnote.[23]

Allenby uses the phrase "integrative cognitivism" to describe the framework in which individual intentions lead to collective effects on a new planetary scale. While the global effect is neither predictable nor teleological, it does reflect human intentionality and the free will of individuals: if we can recognize this fact, we can ascribe intrinsic moral responsibility to human choices. Of course, on a global scale, moral responsibility must be a kind of general, collective human obligation, not something relegated to individuals (those, for instance, who created a certain technology); in many cases, such people – even those acting in good faith – are not in the position to predict possible negative outcomes. Indeed, it is only at the global level that we can acknowledge a kind of emergent collective intentionality operating "through" increasingly complex cognitive systems:

> ... intentionality properly understood arises from the cognitive system(s) within which the individual functions and not from any capacity internal to the individual alone. ... The implication of this discussion for ESEM is clear. We know that we have created an anthropogenetic world and that knowledge creates an overarching moral responsibility to exercise intentionality – design and management – in ways that are ethical and moral.[24]

This means that one has to know "the nature of the complex systems within which one is embedded before one can act with rationality and ethical responsibility to help evolve them."[25]

The concept of so-called integrative cognitivism is linked to the theory of the "actor network" that I illustrated briefly in chapter 1:[26] in keeping with Giddens's and Rowlands's[27] perspectives, Allenby concurs with recent research in cognitive science that indicates that the mind is *extended* in external objects and representations, just as I observed in the

[23] Gorman, 2005.
[24] Ibid., p. 337.
[25] Ibid., p. 339.
[26] Cf. the section "Human and Nonhuman Collectives, Human Agents, and Nonhuman Agents" in chapter 1.
[27] Giddens, 1990; Rowlands, 1999.

previous chapter, where I maintained that our status as "cyborgs" is due in part to the modern human being's extended mind. The mind is extended, as is the intentionality of individuals, which, once it has affected a complex system, can become distorted, complicated, combined, and confused. Integrative cognitivism holds that the mind is always hybrid, as is expressed in an act of cognition during this interplay between internal processes and the external materiality of structures. As we already know, we are cyborgs because of our ancestors' continuous – and evolutionarily successful – off-loading of cognitive functions into the external environment and subsequent re-internalization of those functions when possible and useful.

Adopting this approach helps Allenby to depict the "dynamics of Mind in an anthropogenic earth": technology is the fruit of this continuous cognitive reification that coincides with the evolution of cognition. The three examples of nanotechnology, biotechnology (including genomics), and information technology (the creation of intelligent networks) represent the extension of human design and function into external realms that were previously independent of human cognition. These are examples of "the extension and reification of human Mind in the anthropogenic earth."[28] It is now clear why building an ESEM capability requires an entirely new way of thinking: "It requires that we begin at the very beginning, with an exploration of how we think, and how we know – not just individually but at the level of increasingly broad and complex cognitive systems."[29] Elements of cognition previously considered to be natural have become integrated with human cognition, both individual and collective, and the implications of this combination are difficult to predict.

Further, Allenby strongly emphasizes the fact that in this interplay, individuals' intentions are embedded in complex systems: "... the cognitive process, although still emanating from the individual, becomes different in kind. Cognition and cognitive power become coupled to the particular elements of cultural and technological evolution that characterize different societies; the unpredictable contingency that history so clearly demonstrates is characteristic of human systems."[30] An ESEM capability has to elevate cognition from the implicit and unperceived to the explicit, which will allow us to understand technological processes: once they are organized into an appropriate hierarchy, they can be more readily managed. In so doing, ESEM still requires intentionality and free will and, consequently, "moral responsibility."

[28] Allenby, 2005, p. 327.
[29] Ibid., p. 328.
[30] Ibid., p. 329

When discussing intentionality and free will, Allenby seems to overlook the importance of two ideas:

- Free will is not "performable" only because of external (political) and internal (spiritual) freedom.[31] At a more basic level, free will depends upon the extent of our scientific and ethical knowledge: we can hypothesize a kind of co-evolution between our brains and our ability to produce and disseminate knowledge that furnishes the all-important "elbow room" that, as I demonstrated in the previous chapter, is critically necessary for free will.[32]
- Certainly, intentionality and free will are expressed through and constrained by objective systems that, as integrative cognitivism teaches us, are in turn affected by the freely made decisions of those who use the systems, and these systems are complex and constantly changing. Allenby maintains that individuals' intentionality and free will are the "ultimate source" of the contingencies of human-affected systems – for example, technologies that involve a sort of "derivative intentionality" beyond what was originally planned. The internet is a particularly good example of this: people are the "ultimate source" of this system's potential for good or ill when they use free choice in constructing new websites, for example. Most sites have legitimate uses – to enhance sales of a certain product, for example – but they can also, more or less unexpectedly, serve as vehicles for selling illegal products or cheating people in some way. Following Hurricane Katrina in August 2005, for instance, opportunistic criminals launched several ostensibly charitable websites that in fact routed donations to their own bank accounts rather than to disaster relief organizations. Those who originally designed programs for online fund-raising probably did not anticipate the software's being used in this particular manner.

I think this attitude misrepresents the *autonomous* and *general* role played by external materialities and underestimates their intrinsic features' ability to constrain human cognitive acts through a kind of hybrid interplay between internal and external. As I will more thoroughly explain in chapters 6 and 7,[33] this materiality, when affected by human cognition and action, becomes a *mediator* that can exhibit unexpected moral outcomes, negative, as well as also positive. Moreover, mediators are also related to implicit ways of moral acting,

[31] Ibid., pp. 329–330.

[32] See the sections "Tracking the External World" and "Tracking Human Behavior" in chapter 3.

[33] I have drawn the concept of "moral mediators," which will be illustrated in chapter 6, from what I call "epistemic mediators" that play a role in scientific thinking.

> which I call "templates of moral doing." What I call "morality through doing" is the capacity human beings have to act morally – and implicitly – through doing, without any precise intellectual awareness, as is sometimes the case in the ethics of care. This "acting morally through doing" often consists of building external structures and materialities that can mediate positive moral outcomes, even unexpected ones.[34]

To conclude, I agree with Allenby that the mediated intentionality expressed in large and complex cognitive systems reminds us that we need not evade moral responsibility, and I would further emphasize that circumventing such an obligation arises from a lack of commitment to producing suitable knowledge. As a result, modern people are often jeopardized by the unanticipated global-scale outcomes of localized individual intentions, and this gap between local intentions and global consequences derives from using systems (technologies, for instance) that are just energized by that intentionality itself. Fully understanding our various cognitive systems will help us to avoid damage to the Earth , which is why I strongly support the motto "knowledge as duty."

The Right and the Duty to Information and Knowledge

How is knowledge (and information in general) typically regarded in current ethical debates? Do we already recognize the intrinsic value of knowledge? These are open questions requiring much discussion. Some ethicists contend that an interest in information could indeed have intrinsic moral significance, and they pose questions about a possible fundamental right to information per se that can serve as a foundation for information ethics.[35] The conclusion seems to be that there is not, in fact, a general right to information, that we might have a right only to some kinds of information – religious, for example. In this scenario, the right would derive from an interest in specific content rather than from the content of other rights – for instance, the fundamental right to life and liberty. This would mean that the right to that information is fundamental but not general, so that "providing a comprehensive, plausible theoretical analysis of our information rights will be a complex and nuanced undertaking."[36]

Moreover, there are two kinds of intrinsic value we must consider. The value of being regarded as an end in itself (intrinsic value in the *descriptive*

[34] See also the section "Moral Agents and Moral Patients" in chapter 6; there I will discuss moral mediators and what is called "surrogate intentionality," which is analogous to Allenby's derivative intentionality.

[35] Himma, 2004.

[36] Ibid.

sense) is different from value ascribed because respect is seen as obligatory (intrinsic value in the *normative* sense): unfortunately, assigning intrinsic value to X is neither a necessary nor a sufficient condition for having a fundamental right to X. We might have a right to property, for instance, even if that property lacks intrinsic value. It seems there is no fundamental right to information per se despite the fact that we intrinsically value information.[37]

Anyway, having an informative or cognitive nature is not sufficient for having intrinsic value per se, and, consequently, intrinsic value ascribed in this way to knowledge as duty must be considered only circumstantial. While one day people may intrinsically value all information and knowledge, *we* are not those people. On the basis of such considerations, it seems that there is no fundamental right to information per se despite the fact that we sometimes value information intrinsically.

Information also has an extraordinary instrumental value in promoting the ends of self-realization and autonomy, as we have seen in the previous chapter and as will be described in more detail in the following two. But not all attitudes toward information promote these ends. The questions of whether and how people attribute instrumental value to information are empirical. Usually, it is that part of information we call "knowledge" that endows information with intrinsic value.

Unfortunately, there is no evidence that knowledge is universally regarded positively – some people, for instance, regard higher education as an elitist waste of time and money – and the result is, at best, a nebulous idea of information's moral significance.

Kenneth Himma solves this confusion admirably in a way that is sensitive to the problem of knowledge as duty:

> If it is true that people characteristically value particular kinds of information instead of information *per se*, they are making an objective mistake of some kind: as a normative matter, the mere fact that something counts as information is a good reason to value it; anything that has informative nature *should*, as a matter of objective morality, be valued by rational agents. But the fact that information should be valued this way is enough to justify thinking that we have a general fundamental right to information that ought to be protected by the law in every nation. Morality protects not only those things that we characteristically value, but also those things we *ought to* characteristically value.[38]

This argument posits a sort of intrinsic value that lies somewhere between the two concepts illustrated above. We are concerned with presenting evaluations that people *ought* to make rather than evaluations that

[37] The intrinsic value of information is also maintained by Floridi (2002b): all entities have a minimal moral worth qua information objects and so may deserve to be respected.

[38] Himma, 2004.

people *do* make as matters of empirical fact. It is important for us to acknowledge that some of our interests *must* be protected by moral principles; we cannot merely assume that certain things have moral standing as moral patients with some minimal entitlement of respect. In the present book, many passages containing my suggestions on what people should do are inspired by a dynamic perspective of this nature.

Given these considerations, the argument here differs greatly from the one that infers the existence of various rights from some kind of distinction between intrinsic and instrumental value. In this last case, the argument is instead strongly normative and refers to a clear idea of human well-being and development, not related to or disconnected from people's actual evaluations. People have to assign value to something based on its informative (or knowledge) nature without being swayed by popular opinion or the current and actual average inclinations of "present" human beings. After all, if subsequent forms of morality are the fruit of negotiations among individual human beings and collectives, it seems obvious that a new moral perspective (principle, rule, etc.) results from the general affirmation and acceptance of a particular group's previously normative proposal.

Himma observes that such a claim has no clear foundation in either traditional ethical theory or neoclassical natural law theory. For example, he tells us that John Finnis still justifies the list of human goods based on *empirical* claims about what people of all cultures in all times have almost universally valued.[39] But we can still make sense of our commitment to information and knowledge at an intuitive level. An example related to our discussion of knowledge may be useful here: people ought to value a piece of knowledge either for its capacity to describe in an interesting new way some part of the world or for its ability to detect the unethical outcomes of some technology or technological product that has been discovered, for example, to be a potential carrier of global damage.

To conclude, there is no right to information that is both general and fundamental – there is no fundamental right to information per se. This does not imply, however, that we have no moral rights, general or otherwise, to information. From this ethical perspective, the moral meaning of our motto "knowledge as duty" becomes clear. Knowledge "has to be" a duty, and it relates to specific rights we aim at affirming. We are engaged in assessing extended and circumstantial rights to information and knowledge. The right to attain knowledge in particular and information in general is an important entitlement that I hope will become a universally accepted, objective value.

If we want knowledge to be considered a duty of the supranational society, then the goal should be to generate, distribute, and use knowledge

[39] Finnis, pp. 82–84,

to encourage economic and social development. Doing so, however, demands a great deal more research on creating and constructing knowledge for public and private use and, subsequently, on how to apply it to decision-making activities. The very future of the human species is at issue: if we accept that it is our *duty* to construct new bodies of knowledge, both public and private, we will be better able to avoid compromising the conditions necessary to sustain humanity on Earth.

Unfortunately, at the beginning of the twenty-first century, we can see that in Western European and North American societies, at least, knowledge and culture do not appear to be a priority. Even though there are funds, institutions, and human resources devoted to reproducing cultural activities, the vitality of culture remains in question. For example, we face a real breakdown in many humanistic traditions: in spite of its potential to help, research in humanistic fields – from philosophy to sociology to cognitive science – is voiceless before the problems of society. The intellectual exchange among these disciplines occurs mainly within universities and, as a result, creates a kind of academic short circuit. Intellectuals outside universities are similarly isolated, and their work is rarely woven into the fabric of society in general. It seems that people who do not inhabit intellectual circles value research only if it relates to technology or to some branch of science related to technology, but such knowledge, unhappily, is not sufficient to build a knowledge-based society. Technology, while important, is blind to many problems we will face in the future, and knowledge built only from scientific fields has limited value, as would knowledge drawn only from humanistic fields.[40] Knowledge is a duty, but not just any knowledge: it must be a well-rounded, varied body of information drawn from many disciplines, and even then it must be useful, available, and appropriately applied.

Indeed, we must remember that even in rich countries, knowledge gleaned only from science and technology can be difficult to obtain and exploit. Let us imagine a country with many atomic power plants that require securing. To achieve this goal, highly qualified employees must be produced and maintained, but doing so is contingent upon whether or not collectivities make adequate resources available to ensure the skills

[40] Indeed, both scientific and technological innovations may jeopardize our own safety. In this sense, a precautionary principle is usually called for. However, as the sociologist Frank Furedi (2002) has recently argued, an exaggerated presentation of the destructive side of science and technology may limit people's aspirations and dissipate their potential. Rather than exaggerating the negative side and celebrating safety, as mass media often do, we should acknowledge that it is knowledge and science that give us all the instruments to prevent possible destructive outcomes. As Furedi put it, "it is the extension of human control through social and scientific experimentation and change that has provided societies with greater security than before." On the so called culture of fear, see also Glassner, 1999.

training and general education of suitable experts. Seen in the light of the Soviet Union's decline, the disaster at the Chernobyl nuclear plant is a striking real-world example: it is extremely probable that the nation's dissolution weakened agencies charged with training workers at nuclear sites and with heightening general public awareness of technological and scientific issues. In analogous cases, of course, problems have arisen not only from lapses in training but also from corruption and/or a lack of appropriate ethical knowledge. Moreover, even collectivities with highly qualified personnel and an advanced level of scientific-technological "know-how" can be vulnerable to accidents when that knowledge is used inappropriately, not just when there is a failure of ethical (and, of course, political and legal) controls and knowledge.

Another important issue is related to so-called unintentional power – that is, the power to harm others in ways that are difficult to predict. This problem, which is widespread in computing systems, can certainly result from a lack of knowledge about possible outcomes: "one difficulty with unintentional power ... is that the designers are often removed from situations in which their power (now carried by the software they have designed) has its effect. Software designed in Chicago might be used in Calcutta."[41] The following are eloquent examples of unexpected behavior in computing systems that show – as I stressed at the beginning of this chapter – how knowledge becomes a more critically important *duty* than ever before and must be used to inform new ethics and new behaviors in both public policy and private conduct. Huff eloquently describes some particularly bad cases:

> In the summer of 1991 a major telephone outage occurred in the United States because an error was introduced when three lines of code were changed in a multimillion-line signaling program. Because the three-line change was viewed as insignificant, it was not tested. This type of interruption to a software system is too common. Not merely are systems interrupted but sometimes lives are lost because of software problems. A New Jersey inmate under computer-monitored house arrest removed his electronic anklet. "A computer detected the tampering. However, when it called a second computer to report the incident, the first computer received a busy signal and never called back." ... While free, the escapee committed murder. In another case innocent victims were shot to death by the French police acting on an erroneous computer report. ... In 1986 two cancer patients were killed because of a software error in a computer-controlled X-ray machine.[42]

Finally, a subproblem regarding knowledge as duty concerns problems associated with concrete activity in scientific research. For example,

[41] Huff, 1995, p. 102.
[42] Gotterbarn, 2001.

biomedical research seems so important that it warrants the moral obligation not only to pursue it but also to participate in it, at least when minimally invasive and relatively risk-free procedures are at stake. In this case, the duty concerns individuals who would be required to participate in serious scientific research. The issue is fraught with a variety of moral and legal controversies: pragmatically balancing dangers, risks, and benefits; vigilance against, and the ability to identify, wrongdoing; fully informed consensual participation or possible (justifiable) enforceable obligation; the interplay between human rights and the interests of society; the possibility of industrial profits from moral commitment; the potential exploitation of poor people; the role of undue inducements; the special case of children, and so on.[43] When one participates as a subject in a scientific experiment, the commitment to scientific knowledge is not necessarily derived from the idea that knowledge is a good in itself, but rather from the related duty of beneficence, our basic obligation to help people in need.

RATIONAL KNOWLEDGE AND VALUES

The rise of modern science has generated an increase in *rational* knowledge. In the last century, especially in Western Europe and the United States, the fields of philosophy, logic, epistemology, and cognitive science have effectively mapped out the rational capacities of both human beings and their technological products. But a similar advance has not been made in ethics or in the ethical aspects of public policies and decision making; we have not yet explored deeply enough the relationship between rationality and *values*. The "construction" of a modern, ethical, self-correcting knowledge is far from being achieved, yet it is critically important if we are to balance the needs of the poor and the rich around the world. In Europe, certain historicist and nonrational traditions have almost "culturally" killed the value of rationality derived from Galilean science and have generated various kinds of sterile (and dangerous) irrationalism. In the United States, overestimating the importance of productivity and technology is leading to a national collective consciousness that is often devoid of cultural awareness: only technological products seem to have any effect on subculture and subcultural identity. For example, the habits of everyday people have been shaped by corporate-oriented mass media advertising and the stereotypes it engenders. Even those messages that stress modern rationality are undermined by their triviality; this fact has also favored the growth of other kinds of regressive identification, like those that involve medieval-style peudo-religious convictions.

[43] Harris, 2005.

We must instead use the current findings of philosophical, epistemological, cognitive, and human sciences – findings that support the idea of knowledge as duty – to build new kinds of *rational traditions*. I think these traditions can be enhanced by reestablishing knowledge as a priority and then using it to guide social choices appropriately. In this way, we will also be able to recover and restore some of the treasures of the Western world's cultural and scientific history.

It is worth noting that some transnational communities, those composed of "a network of professionals with recognized expertise and competence in a particular domain and an authoritative claim to policy relevant knowledge within the domain or issue area," are called *epistemic* communities.[44] Burkard Holzner and John Marx, who introduced the term, define these entities as "knowledge-oriented work communities in which standard cultural and social arrangements interpenetrate around a primary commitment to epistemic criteria in knowledge production and application."[45] Holzner and Marx have received attention in the recent literature on international cooperation for their work on the management of the "commons." For example, they describe epistemic communities that exist among Antarctic scientists, international marine lawyers, and radio spectrum and telecommunications engineers.[46]

KNOWLEDGE AS DUTY AND THE NEED FOR TRANSDISCIPLINARITY

Of course, if knowledge is to be considered a duty, the cognitive, logical, methodological, and epistemological problems of distributing it are crucial. The fundamental problems of knowledge dissemination will be addressed in the last two chapters, but some introductory issues are presented here, particularly regarding the importance of interdisciplinary efforts: the present book itself aims at being an interdisciplinary product. I think this methodological and epistemological commitment is vital because it is part of the idea of knowledge as duty and is therefore relevant to the concerns of traditional moral philosophy. The issues are general, and many of them are very much open to future research. They primarily concern the importance of increasing knowledge about creative and manipulative reasoning, and about the role in ethics of externalities and moral mediators. They also include the challenges of distributing scientific and ethical knowledge – from the problem of unexpressed and super-expressed knowledge and knowledge management to the so-called new knowledge communities.

[44] Haas, 1990.
[45] Holzner and Marx, 1979, p. 108.
[46] Vogler, 2000, p. 28.

Creativity and Model-Based Knowledge

Some transdisciplinary subtopics affect the central aspects of "knowledge as a duty" in multiple ways. Thanks to clarification provided by cognitive science and artificial intelligence, epistemological studies on creative and explanatory reasoning can bolster the argument for regarding knowledge as duty. For example, the study of abductive reasoning and model-based reasoning can be exploited to stress and elicit knowledge about particular issues in scientific and ethical reasoning. We have indeed been able to increase knowledge about creative reasoning (for instance, in science) and diagnostic reasoning (for instance, in medicine) (two types of abduction, cf. chapter 7). Moreover, the mechanisms of many kinds of model-based reasoning – thinking with diagrams, working through simulations and thought experiments, and manipulating material models, for example – are now clear. These achievements are important because they can rationally illustrate the many methods available for solving problems in the best way.

Many topics in this book relate to the importance of creativity: knowledge plays a creative role in attributing new values to external things (see chapter 1, the section "Women, Animals, and Plants"). Further, in chapter 6 I highlight some of the research in cognitive science and artificial intelligence that affects ethical knowledge and reasoning. (Moreover, some sections in the second half of chapter 7 use a basic epistemological approach to explore how abduction is an important form of explanatory reasoning, what factors serve as "reasons" in decision making, and the nature of model-based reasoning and epistemic mediators.)

Building a rich body of ethical knowledge that addresses today's problems is certainly a duty, just as it is to develop knowledge in other areas like science and medicine (see the sections "Good Reasons and Good Arguments" and "Creating and Selecting Reasons" in chapter 6). Enhancing our ethical wisdom requires work on several fronts: developing verbal/propositional and model-based knowledge (see the section "Model-Based Moral Reasoning" in chapter 6); exploring manipulative and "through doing" knowledge (see "Being Moral through Doing" in chapter 6); and taking advantage of recent epistemological research and studies in the areas of cognitive science and artificial intelligence. And there are still more challenges created by the idea of "knowledge as duty": dealing with the intrinsic "incompleteness" of various ethical and epistemological settings, for ethical reasoning always occurs in the presence of incomplete information (see "Expected Consequences and Incomplete Information" in chapter 6); the effects produced by inconsistencies (what can we do when we are faced with ethical issues that present contradictory aspects?); and the problem of comparing alternatives in theoretical and deliberative settings (see "Comparing Alternatives and Governing Inconsistencies" in chapter 6).

Knowledge and Moral Mediators in Computational Philosophy: Distributing Morality

Some external objects and structures that we use in science, those to which we delegate cognitive aspects and roles, I call *epistemic mediators* – a blackboard with a diagram, for example (cf. chapter 7 of this book). In a recent book on creative reasoning, I described epistemic mediators not only as external objects and structures but also as human organizations that distribute externalized cognitive potentialities.[47] Cognitive mediators function as enormous new external sources of information and knowledge, and, therefore, they offer ways of managing objects and information that cannot be immediately represented or found internally using only "mental" resources. Analyzing these external structures is especially important in clarifying the role of the media and of computational and information techniques. Epistemic mediators also help to organize social and cognitive decisions made in academic settings: they may be artifacts in a scientific laboratory, for example (a telescope or a magnetic resonance imaging machine), or even the collective of scientists itself, which is characterized by a specific distribution of cognitive roles, skills, and duties.

I think the best approach to studying these problems is to use so-called *computational philosophy*. The advent of certain machines and various rational methods and models brought about a computational turn during the last century, and this shift has revealed new ways to increase knowledge by embedding it in scientific and technological environments and by reshaping its traditional major topics.[48] Just to give an example, PCs and the internet have an important role in improving scientific research. In the new century, computational philosophy will allow an analysis of recent problems in logical, epistemological, and cognitive aspects of modeling activities employed in scientific and technological discovery. Computational philosophy supplies modern tools (new concepts, methods, computational programs and devices, logical models, etc.) that can be used to reframe many kinds of cultural (philosophical, ethical, artistic, etc.) knowledge that would remain inaccessible using old approaches that rely mainly on the use of mere "narratives."

It is in this intellectual light that I introduce the concept of the *moral mediator*. Moral mediators play an important role in reshaping the ethical worth of human beings and collectives and, at the same time, facilitate a

[47] Magnani, 2001a.
[48] Magnani, 1997.

continuous reconfiguration of social orders geared toward rebuilding new moral perspectives. To revisit an idea from the subsection "Ends Justify Means" in chapter 1, the emotional impact of moral mediators like ecotage and monkey-wrenching is an example of their power and is evidence that moral mediators can be vehicles that allow us to obtain otherwise unavailable ethical information and values. A full description of this new concept appears in chapter 6 in the section "Being Moral through Doing: Taking Care."

Finally, thinking in terms of cognitive capacities, a human being can be considered a kind of "thing" that can incorporate information, knowledge, know-how, cultural tradition, and so on, just as cognitive objects like books, PCs, and sculptures do. Unfortunately, human beings are sometimes assigned less value than things – remember the library book discussed in chapter 1: moral mediators can help us to redefine people as worthy of new moral consideration (see the chapter 1 subsection "The Thing-Person," as well as chapter 3).

Unexpressed Knowledge, Super-Expressed Knowledge, Responsibility

As stressed by George Von Krogh, Kazuo Ichijo, and Ikujiro Nonaka, the need to elicit and distribute *unexpressed knowledge* – that is, isolated pockets of "hidden" information in an organization – presents a considerable challenge in knowledge management (KM), one that must be addressed as we take on the knowledge-as-duty problem.[49] KM helps to generate assets for an organization by transferring knowledge and expertise, allowing them to be more effectively utilized; it leverages an organization's *intellectual capital* (people's knowledge and expertise) and, in combination with accurate, qualitative, and pertinent data, helps to provide an improved product. In KM, it is clear that what people know and can learn is more valuable than any particularly important in the context of the opposite problem, what I call *super-expressed* knowledge (and information): a huge, expensive cache of available knowledge that is overused, unused, and/or not useful, such as the mountains of explicit statistical data that rich societies constantly produce but never use or use only as a curiosity.

Super-expressed knowledge is also a problem when it comes to the vast collection of knowledge on the internet and the ethical question of privacy (see the dicussion later in this chapter). I maintain that while the idea of "knowledge as duty" certainly applies to many scientific, social, and political problems, it is an even greater factor in mapping ethical

[49] Von Krogh, Ichijo, and Nonaka, 2000.

behavior in this new cyberage. The internet provides a staggering amount of information that can influence human beings' welfare, and if we fully and universally acknowledge the internet's immense power, we can immediately see that moral problems arise: issues of accuracy, deception, copyright, patents, trademark abuse, and retention of information, just to name a few, have become much more complex as a result of the internet. While these problems are vexing, I contend that "knowledge as duty" will help to guide us through this ethical labyrinth and, for example, reinforce the right to privacy.

This book also reconsiders the problem of human dignity in the era of globalization by analyzing mechanisms used to form rational ethical arguments (chapters 6 and 7) and by providing integrated analyses of current debates: cloning (chapter 2), cyberprivacy (this chapter), hybrid people and consciousness (chapter 3), and the environment (chapter 1). Two common denominators of all of these issues are an absence of appropriate and situated knowledge and, as a result, widespread cultural wariness and skepticism. Knowledge is often abstract, synthetic, and general: cloning is just treating people as means; privacy is needed just to preserve identity; preserving nature is worthwhile only to the extent that it serves human interests. The problem is that by practicing cloning, someone can be considered to be treating people as means and, at the same time, as providing a necessary and useful service in particular cases. Preserving nature can be deemed too dangerous to people in certain settings, and excess privacy might threaten people's affective and social lives. This skepticism clearly shows the devastating problems that can arise when our moral deliberations are influenced by too-general perspectives that either ignore the concrete details of specific situations or draw on extra-moral reasons because the available moral options are hopelessly vague and abstract. Furthermore, a lack of knowledge makes it almost impossible to behave responsibly in any situation.

As stated earlier, human dignity is inextricably linked to the concepts of responsibility and freedom, and in chapter 5 the idea of "knowledge as duty" will be explored more fully by analyzing bad faith and its impact on human beings' control over their own destinies. I contend that both moral knowledge and scientific knowledge – when available, exploitable, and appropriate – are powerful factors in developing responsibility and protecting freedom; the moral gap caused by bad faith results mainly from inadequate and "incomplete" knowledge and information. Enriching our knowledge about "our condition" is critically important if we are to fortify human freedom and create incentives to adopt more responsibility.

New "Knowledge Communities": Humans as Knowledge Carriers

Technology is central to modern society, but those who create it are not taught to anticipate or investigate its impact, and, unfortunately, those who focus on technology's influence rarely understand the technologies themselves. I think that universities have a special responsibility, and therefore a unique opportunity, to address this issue. Especially since the nineteenth century, European and North American universities and technical institutions have constructed a markedly bifurcated curriculum: one track is a technical/engineering/scientific curriculum, with courses in math, computing, and engineering taught in one intellectual region of the campus, while courses in the social sciences, arts, and humanities courses are taught in some other area. But the central questions of society are not about engineering or technology as such – rather, they encompass social, economic, political, and cultural issues. Taking on these questions will require the coordination of university curricula, a fruitful integration of engineering and liberal arts that might pair, say, bioprosthetics and disability literature. Also necessary is a partnership between research and service, and in this light the relationship between universities and industry is important. New *knowledge communities* have to be built. We must envision new alliances among universities, institutes, and businesses and redefine our very idea of "knowledge" itself and how it is distributed. For example, we must forge creative new roles for the information found in books, journals, and databases and then implement the newly obtained data in society so that we can reshape information security, patent rights, licensing rights, and so on.[50]

While technology provides benefits like higher living standards, greater choice, more leisure time, and improved communication, it also brings with it both environmental and human risks. Technology alienates people from nature, concentrates political and industrial power in dangerous ways, and begets industrial nations that are capital-intensive rather than labor-intensive; the result is increased unemployment and an overdependence on ostensibly neutral experts who in fact have interests and agendas of their own.

Though valuable in many ways, technology is not always benign, and it has the capacity to cause many kinds of damage. It can lead to societal uniformity, impersonal treatment, the manipulation of certain groups of people, and too-narrow specialization. Another negative outcome is uncontrollability, a situation where "separate" technologies from an interlocking system constitute a total, mutually reinforcing network that seems to lead a life of its own, self-perpetuating and all-pervasive. Finally,

[50] Cf. also the earlier discussion of ESEMs and "trading zones" in the section "The Gap between Individual and Collective Intentionality."

there is also the risk of what is, from a Marxian perspective, the classic negative consequence: the alienation of the worker. Intellectuals have offered many antidotes to technology's ill effects: embracing a new ethics of responsibility; returning to the genuine engagement we experience when we focus on simple things rather than on production and consumption; recovering an imaginative and emotional life as well as religious commitment; redirecting technology according to social mores or religious constraints; and, finally, using new political projects to reestablish the importance of the personal and the individual as opposed to the materialistic and the impersonal.

I think the role of human beings as *knowledge carriers* – as "biological" knowledge repositories, processors, disseminators, and users – is pivotal in our era of technology and globalization, an idea that I treated in the last section of chapter 1 and that I will treat in the section "Globalization, Commodification, and War" in chapter 5. In the globalized world, knowledge-carrying human beings enjoy increased status in various work collectives, social settings, and environments that have become more information-intensive, even if they are also increasingly disenfranchised and fragmented. Such people collaborate with (and, in some sense, are integrated with) nonhuman objects that manage an enormous quantity of information. There are certainly many perils lurking in this globalized era – the alienating and exploiting of certain vulnerable groups of people comes to mind, as does the obliteration of local cultures. But I think these new trends in technology also have the potential to benefit humankind in many ways: disseminating and universalizing information and eliciting unexpressed knowledge could bring exciting new possibilities and new opportunities for freedom and free choice.

Consequently, human beings, endowed by technology with new social intelligence, will be able to draw from their own reservoirs of knowledge and, at the same time, effectively wield new technology and artifacts in a way that heightens their intellectual and cooperative powers. As knowledge carriers, people are just as important as the nonhuman artifacts to which they are so closely linked; as we saw in the previous chapter, the human hybrid is, in certain respects, greater than the sum of its parts. Looking at human knowledge carriers in this way, it is easy to respect them as much as we respect, say, a computer, and this attitude will facilitate a critical shift in perspective as we face the important task of learning to "respect people as things."

IDENTITY AND PRIVACY IN THE CYBERAGE

Even though producing knowledge is an important goal, actually doing so in certain circumstances is not always a welcome prospect: we must be cautious when dealing with issues of identity and cyberprivacy, where the excessive production and dissemination of knowledge can be dangerous.

The ethical problem of privacy is related to the theoretical problem of identity. Neuroscientists have discovered that some modular processors do not extract processes from the environment but, rather, from the subject's own body and brain. The brain contains multiple representations of itself and of the body, such as the body's physical location in space, thanks to continuously updated somatic, kinesthetic, and motor maps.[51] The brain also contains the representation of identity, a narrative autobiography encoded in episodic memory and higher cognitive functions like action, perception, verbal reasoning, and a "theory of mind" that help us to interpret other people's behavior. It can be hypothesized that all these modules operating simultaneously in the conscious workspace accounts for the subjectivity of the self and its own identity. Once available in the conscious workspace, these modules' activity could be inspected by other processes, thus setting the stage for reflexive and higher-order consciousness.[52]

Hence, it seems that our consciousness is formed in part by representations of ourselves that can be stored outside the body (narratives that authenticate our identity, various kinds of data), but much of consciousness also consists of internal representations of ourselves (a representation of the body and a narrative that constructs our identity) to which other people do not have access. We rarely transfer this second type of representation onto external mediators, as we might with written narratives, concrete drawings, computational data, and other configurations. Such internal representations can be also considered a kind of information property (moral capital) that cannot be disseminated – even to spouses, since the need for privacy exists even between husband and wife. There are various reasons for withholding such representations: to avoid harm of some kind, for example, or to preserve intimacy by sharing one's secrets only in love or friendship. But perhaps the most deleterious effect of the loss of privacy is its impact on free choice. The right to privacy is also related to respect for others' agency, for their status as "choosers"; indeed, to respect people is also to concede that it is necessary to take into account how one's own decisions may affect the enterprises of others.

When we lose ownership of our actions, we lose responsibility for them, given the fact that collective moral behavior – as I contend – guarantees the ownership of our own destinies and to the right to choose in our own way. When too much of our identity and data is externalized

[51] Damasio, 1999.
[52] Cf. Fletcher et al., 1999; Gallese et al., 1996; and Weiskranz, 1997, quoted in Dehaene and Naccache, 2001, p. 31.

in distant, objective "things," the possibility of being reified in the externalized data increases, and our dignity subsequently decreases. By protecting privacy, we protect people's ability to develop and realize projects in their own way. Even if unlicensed scrutiny does not cause any direct damage, it can be an intrusive, dehumanizing force that fosters resentment in those who are subjected to it. We must be able to ascribe agency to people – actual or potential – in order to conceive of them as fully human. It is only when our freedom is respected and we are guaranteed the chance to assume responsibility that we can control our lives and obtain what we deserve.

New Moral Ontologies

Cyberwarfare, cyberterrorism, identity theft, attacks on abortion providers, Holocaust revisionism, racist hate speech, organized crime, child pornography, and hacktivism – a blend of hacking and activism against companies through their web sites – are very common on the internet.[53] As a virtually uncontrollable medium that cuts across different sovereign countries, cyberspace has proved to be fertile ground for new moral problems and new opportunities for wrongdoing, mainly because it has created new moral "ontologies" that affect human behavior.

Much more than in the past, information about human beings is now being simulated, duplicated, and replaced in an external environment. Beyond supports of paper, telephone, and media, many human interactions are strongly mediated and potentially recorded through the internet. At present, identity must be considered in a broad sense: the amount of data about us as individuals is enormous, and it is all stored in external things/means. This repository of electronic data for every individual human, at least those in rich countries, can be described as an external "data shadow" that, together with the biological body, forms a kind of cyborg that identifies or potentially identifies him or her, The expression "data shadow" was coined in Sweden in the early 1970s: it refers to a true, impartial image of a person from a given perspective.[54] What will happen when a silicon chip transponder, once surgically implanted in a human being, can transmit identifying data like one's name and location through a radio signal? Of the many, many possibilities, one of the more benign is that it may render physical credit cards obsolete. And what will be the result when a transponder directly linked with neural fibers in, say, one's arms can be used not only to generate remotely controlled movement for people with nerve damage, but also to enhance the strength of an

[53] Van den Hoven, 2000, p. 127.

[54] Gotterbarn, 2000, p. 216. Cf. also the amazing example of the "pizza profile" in the pizza parlor caller ID, illustrated in Moor, 1997.

uninjured person? For the blind, extrasensory input will merely be compensatory data, but for those with normal sight it will be a powerful "prosthesis" that turns them into "super-cyborgs" with tremendously powerful sensory capabilities. What about the prospect of "hooking up" a nervous system to the internet? It will revolutionize medicine if it becomes possible to download electronic signals that can replace disease-causing chemical signals. It is worth remembering that the human nervous system is, by nature, electrochemical – part electronic, part chemical.[55]

When delegating information or tracing identity, human beings have always used external mediators such as pictures, documents, various types of publications, fingerprints, and so on, but those tools were not simultaneously electronic, dynamic, and global, as is the case with the internet. Where we exist cybernetically is no longer simple to define, and new technologies of telepresence continue to complicate the issue.

NEW HUMAN BEINGS: BIOLOGICALLY LOCAL AND CYBERNETICALLY GLOBAL

This complex new "information being" involves new moral ontologies. We can no longer apply old moral rules and old-fashioned arguments to beings that are both biological (concrete) and virtual, situated in a three-dimensional local space but potentially existing also as "globally omnipresent" information packets.[56] It is easy to materialize cybernetic aspects of people in three-dimensional local space, even if it is not quite as spectacular as it was in the old *Star Trek* episodes, when Scottie beamed the glimmering forms of Captain Kirk and Mr. Spock from the transporter room to the surface of an alien planet. Human rights and the notions of care, obligations, and duty have to be updated for these cybernetic human beings. These notions are usually transparent when applied to macro-physical objects in local spatiotemporal coordinates. They suddenly become obscure, however, in cyberspace:

> What is an agent? What is an object? What is an *integrated product*? Is that one thing or two sold together? What is the *corpus delicti*? If someone sends an e-mail with sexual innuendo, where and when does the sexual harassment take place? Where and when is it read by the envisaged addressee? Or when it was typed? Stored on the server? Or as it was piped through all the countries through which it was routed? And what if it was a forwarded message?[57]

[55] Cf. Warwick, 2003, p. 135.
[56] Schiller, 2000.
[57] van den Hoven, 2000, p. 131.

As I will illustrate in the next section, new technologies threaten to increase our moral ignorance. I have previously pointed out that the idea of "knowledge as duty" is relevant to scientific, social, and political problems, but it is even more important when we are faced with reinterpreting and updating the ethical intelligibility of behaviors and events. Clearly, the problem of moral epistemology must be addressed.

The new moral issues of the cyberage are well known. Certain offenses, like financial deception and child abuse, have been perpetrated for years via traditional media, and the problem has only been exacerbated by the internet. Others issues have arisen as a result of new technology – problems like hacking, cracking, and cooking up computer viruses, and the difficulties created by artificial intelligence artifacts, which are able to reason on their own.[58] Also needing attention is the proliferation of spam. The utility of spamming clashes with the time spent handling data trash and garbage by the recipient: in this sense, it can be said that spamming externalizes costs.

Yet another challenge is the huge problem of the digital divide. While millions of people are active cyborgs, a gulf has opened between them and others numbering in the billions. Members of the latter group come from both rich and poor countries, and they have not attained cyborg status. A further split, at least inside the "cyborg community," has occurred because of the way technology dictates the speed, ease, and scope of communication; the dominance of certain text features creates an unlevel playing field, and the disparity between computer users with the most sophisticated equipment and those with more modest systems indicates that we are far from the ideal of an unmediated internet. And the list continues: professional abuses, policy vacuums, intellectual property questions, espionage, a reduced role for humans – all these things create further moral tangles. Finally, if we consider the internet as a global collection of copying machines, as Mike Godwin contends, huge copyright problems arise.[59]

In light of the motto "knowledge as duty," the internet makes available many goods that are critical for human welfare by increasing access to information and knowledge regarding people's biological needs: food, drink, shelter, mobility, and sexuality. It also serves their rational needs

[58] Steinhart, 1999. See also chapter 3 of this book.

[59] Godwin, 1998. See also the classic Johnson, 1991, and Stallman, 1991. The range of ethical problems that have been generated by computer ethics is so large that some authors have maintained that it is a "unique" field of applied ethics in the sense that its ethical issues might happen to be unique (on the "uniqueness debate," see. Tavani, 2002). Certainly, computer ethics is a wonderful example of moral progress, in terms of both new moral issues/objects and new moral principles.

by providing information, stimulating imagination, challenging reason, explaining science, and supplying humor, to cite a few examples.[60] If the internet's power, fecundity, speed, and efficiency in distributing information are universally acknowledged, moral problems immediately arise, related to distortion, deception, inaccuracy of data, copyright, patents, and trademark abuse.[61] It also seems that the internet facilitates bureaucratic control: in Africa, for instance, it increases control over populations treated as the passive recipients of state policies.

We must safeguard freedom of information and, at the same time, preserve privacy, the right to private property, freedom of conscience, individual autonomy, and self-determination. Techno-libertarians say that VR (virtual reality) rapists (that is, men who simulate rape in virtual reality systems) are "of course assholes, but the presence of assholes on the system is a technical inevitability, like the noise in a phone line."[62]

Even in the complicated world of cyberspace, however, Mill's principle still establishes the moral threshold: only the prevention of harm to others justifies the limitation of freedom.[63] Mill also observes that truth emerges more easily in an environment in which ideas are freely and forcefully discussed. It is well known that the Communications Decency Act passed by the United States Congress in 1996 had a very short life and was quickly struck down by the U.S. Supreme Court.[64]

Cyberprivacy

We have said that the post-human "cyborg" comprises a vast quantity of external information and that it is possible to "materialize" many of our cybernetic aspects – at home, I can, in principle, print out a picture of you as well as gather information about your sexual, political, and literary preferences based on records of your online purchases. Consequently, those aspects of private life are not available just to the person they concern; because they are potentially "globally omnipresent," everyone can, in theory, see them. We must note that computer technology also distances the human body from the actions performed, and because it is

[60] van den Hoven, 2000.

[61] Goldman, 1992; Mawhood and Tysver, 2000. On the internet's retention of information by corporations, governments, and other institutions, see Davis and Splichal, 2000.

[62] Dibbell, 1993. Some recent perspectives on the ethics of the "meat-free existence" in cyberfeminism and on the problem of "women and technology" are given in Flanagan and Booth, 2002, and in Adam, 2002.

[63] Cf. chapter 1, the section "Respecting Things as People."

[64] On the destiny of free speech in the digital age, see Godwin, 1998. Many data on the problem of the digital divide in the United States and Europe are given in Compaine, 2001.

an automating technology, it has a great potential to displace human presence.

The problem of privacy is related to the so-called Panopticon effect. The Panopticon, conceived by Jeremy Bentham in 1786, was an architectural design for prisons that allowed very few guards to keep an eye on a great many convicts. The structure was a round, multistory building with each floor's cells arranged in a circle around a hollow core. Rising from ground level in the center, in a sort of round atrium, was a guard tower with windows facing out toward the cells. Because each cell was illuminated by a window to the outside, and the cell's inner wall consisted only of bars, guards could easily monitor the prisoners, every one of whom was completely visible at all times. Michel Foucault described the Panopticon as a metaphor for the "mechanisms of large-scale social control that characterize the modern world."[65] Of course, a person's being on display is enough by itself to effect social control: many psychological studies have observed that when people believe they are visible to others, they behave differently than they do when they believe they are out of view. Foucault considered the practices of medicine, psychology, and sex education to be forms of social control "because they create a world in which the details of our lives become symptoms exposed to a clinical *gaze* – even if not actually looking."[66]

The internet can be seen as a modern Panopticon that makes all of us visible to anyone in a position to see. There is a tremendous amount of information on the net: on display are census data, addresses, credit card purchases, credit histories, medical information, employment histories, evidence of intimate relationships, indicators of religious inclinations – the list could go on and on. But unlike Bentham's prison, where prisoners could be watched only from a single point – the central guard tower – we can be "viewed" from billions of sites, for all it takes is a computer with an internet connection. Computer networks have no centers, and almost any portion can operate as an autonomous whole, which is also what makes controlling the network so difficult. As I have already illustrated, a detailed computational "shadow" of a private life can easily be built by collecting information that is available online.

The availability of too much personal information increases our vulnerability to all kinds of dangers, so protective measures must be instituted; however, monitoring the communications of others contradicts the universal right to privacy. Although there are some cases in which too much privacy can be catastrophic – when families hide the abuse of women and children, for instance – the cyberage renders privacy increasingly important. Heightened privacy poses its own difficulties, of

[65] Foucault, 1979; Reiman, 1995.
[66] Reiman, 1995, p. 29.

course, bringing challenges that include greater costs as well as many new needs – the need for developing new ethical and technical knowledge, establishing new political priorities, and passing new laws.

There are those who maintain that the more privacy we have, the harder it is for society to gather the information needed to apprehend and punish criminals, although I am skeptical about that assertion. Members of society must be protected not only from being seen but also from feeling visible, which allows us to see privacy both as a basic right of which some people are deprived and, even more important, as a method of controlling who may have access to us.[67] Some ethicists contend that this second idea of privacy leads to anomalies: for example, the moral right to privacy has to be limited to "my" right that others be deprived of their access to me. Nevertheless, we certainly want to establish laws that protect privacy and guarantee its status as both a moral and a legal right, even (and especially) in the era of the so-called Information Panopticon.[68]

The internet and the so-called Intelligent Vehicle Highway Systems (IVHS) have created "external" devices that challenge our established rights to privacy, making it necessary to clearly define and shape new moral ontologies. Recently, ECHELON, the most powerful intelligence-gathering organization in the world, has garnered attention; it has been suspected of conducting global electronic surveillance – mainly of traffic to and from North America – that could threaten the privacy of people all over the world. The European Union parliament recently established a temporary committee on the ECHELON interception system (1999–2004) to investigate its actions.[69]

Because so much information has become available about so many people, we must actively attempt to anticipate and avoid possible negative consequences. The ostracism and stigmatization of minorities could increase and generate an explicit loss of freedom. Moreover, choice may be restricted because of an implicit loss of freedom: for example, technology may one day lead a person to say, "I am no longer free simply to drive to X location *without leaving a record*," an idea I will explore further. Less privacy could also mean the loss of protection against insults that "slight the individual's ownership of himself" (as happens in the case of prisoners or slaves): consequently, people will say *mine* with less authority and *yours* with less respect. Indeed, when we lose ownership of our actions, we lose responsibility for them, given the fact that morality – as

[67] Fried, 1984.
[68] Ibid., pp. 32–34; see also Edgar, 1997, and Tribe, 2000.
[69] The 2001 draft report "on the existence of a global system for the interception of private and commercial communications (ECHELON interception system)" can be found at <http://www.europarl.eu.int/tempcom/echelon/pdf/prechelon_en.pdf>.

I contend – involves the right to own our own destinies and to choose in our own way.[70] Finally, the risk of infantilizing people increases, as does the likelihood of rendering them conformist and conventional, and they are, as a result, faced with a greater potential for being oppressed. Given these possibilities, it is apparent that privacy is politically linked in many ways to the concepts of liberalism and democracy.[71]

Philosophy's notion of the fundamental elements of privacy is clearly evoked by John Stuart Mill: "The only part of the conduct of anyone, for which he is amenable to society, is that which concerns others. In the part which merely concerns himself, his independence is, of right, absolute. Over himself, over his own body and mind, the individual is sovereign."[72]

Most authors agree that a cyberdemocracy must be established to emphasize the role of civil rights and responsibility in cyberspace. Johnson's argument about the role of internet technology in promoting democracy concludes in an enigmatic way: the internet "has the capacity for hegemonic as well as democratic relationships." She goes on to pose many provocative questions: (1) How is internet development related to the digital divide? (2) Does the internet move toward private interests and market focus? (3) Can the participation it guarantees influence decision-making processes? (4) Does the internet threaten the autonomy and democratic status of nations? (5) To what extent is there privacy on the internet, and to what extent is it a medium of surveillance? It seems the internet is in the service neither of enhancing democracy nor of promoting global democracy.[73]

Finally, we can distinguish between two kinds of privacy violations: the first involves disseminating intimate information to a general audience without the consent of the person involved,[74] and the second occurs when such personal information is used to make decisions about the individuals

[70] In the following section, I will describe the problem of the ownership of our destinies and of freedom and responsibility. Cf. also chapters 3 and 5.

[71] Reiman, 1995, pp. 35–44.

[72] Mill, 1966, p, 14. Other aspects and benefits of privacy, like promoting mental health and autonomy, are illustrated in Gavison, 1984.

[73] Johnson, 2000, pp. 190–196. Recently, information ethics scholars have increased knowledge about many of these and other topics: see, for example, the classic Maner, 1980; Moor, 1985; Gotterbarn, 1991; Rogerson, 1996; and Johnson, 1994; more recently, see Floridi, 1999, 2002a, 2004; Bynum, 2004, 2005 (on Norbert Wiener's vision of the information society); Floridi and Sanders, 2003, 2004; Moor and Bynum, 2002; and Bynum and Rogerson, 2004.

[74] Elgesem, 1996. On the role of cookies, sniffers, firewalls, data encryption, filtering systems, digital signatures, authentication, and other technical tools that may either threaten or protect privacy, see Beckett, 2000; Weckert, 2000; and Spinello, 2000. EU countries introduced Data Acts to protect data concerning people in 1973 and 1974. The U.S. government prefers to promote self-monitoring and self-regulation.

involved without their knowledge, possibly to harm them. There is always a conflict between privacy and freedom of information. Medical information, while rarely disseminated, can become emotionally charged, for a person may feel lost, unprotected, and vulnerable if the information is in fact spread. Medical records may be communicated to others in certain circumstances, as in the case of epidemiological research. Even in this case, however, each participant's privacy can be protected by assigning them pseudonyms.

To conclude, it really does seem that privacy can be considered both an intrinsic and an instrumental value that is essential to democracy, autonomy, and liberty.[75]

Privacy, Intimacy, Self-Deception, and the Ownership of Our Own Destinies

Inadequate privacy surely injures self-esteem and assaults the feeling of human dignity whenever this lack is a clear threat to individual liberty.[76] We have said that the post-human "cyborgs" are designated as such because they are composed of a huge quantity of electronically incorporated external information about their identities. Externalizing information about ourselves in "things" that are just "over there," which, of course, makes that information available to others, puts in jeopardy very important ends and relations like respect, love, friendship, and trust.[77] These concepts, through which we express some of the more sophisticated aspects of "human animals," are simply inconceivable without privacy.[78]

If we consider privacy to be the ability to control information and knowledge about ourselves, it is obvious – as illustrated by the Panopticon effect – that we are less free when we are observed or when we believe we are being observed. Electronic monitoring drastically reduces the power we have over information about ourselves: "monitoring obviously presents vast opportunities for malice and misunderstanding on the part of the authorized personnel. For that reason the subject has reason to be constantly apprehensive and inhibited in what he does."[79]

75 See also Tavani, 2000; Johnson, 1994.

76 Bloustein, 1984, p. 160.

77 An example of externalizing information is provided by van Wel and Royakkers (2004), who investigate the possible negative impact on privacy of web data mining. Web data mining concerns turning data into valuable information, which may contribute to harming our privacy.

78 Fried, 1984, p. 205.

79 Ibid., p. 216.

Moreover, because intimacy is the conscious sharing of information about ourselves that is typically not revealed to others, privacy is the means to create what has been called the "moral capital" that we spend, for example, in love and friendship. Privacy grants control over information so that we can preserve intimacy. Even between husband and wife, the restraints of privacy hold; although marriage implies a voluntary renunciation of private information, one spouse never knows what the other has chosen not to share.

If we consider trust to be the expectation that another person will behave morally, privacy becomes essential: there cannot be trust when there is no possibility for error. To foster trust, not all information need be shared: a person cannot know that she is trusted unless she has the right to act without constant surveillance; only then can she freely choose to honor or betray the trust, and it is privacy that furnishes that fundamental right.

So in order to engage in intimate relations like love, friendship, and trust, we must reveal some of ourselves; but if this information is seen by too many others, it loses its intimate nature so that love, friendship, and trust are threatened. It now appears clear that having no control over information about ourselves has a deleterious effect on love, friendship, and trust and is, consequently, an assault on our liberty, personality, and self-respect.

Something more must be said about the threats posed by the loss of privacy. The right to privacy also relates to the idea of respecting other people as "choosers": "To *respect* someone as a person is to concede that one ought to take account of the way in which his enterprise might be affected by one's own decisions. By the principle of respect for persons, then, I mean the principle that every human being, insofar as he is a qualified person, is entitled to this minimal degree of consideration."[80]

Consequently, privacy can be defined as protection against interference with a person's way of realizing and developing her interests, in terms both of practical objectives and of merely shaping one's own identity over time. By protecting privacy, we preserve the ability to develop and realize projects in our own ways. If respect for people is related to respect for their rights, privacy becomes paramount. A monitored individual is indeed inclined to "give himself away,"[81] and so he is compelled to appear false, cold, and unnatural. This can affect the attitude of other people whose esteem, love, and friendship he desires. This kind of humiliating self-betrayal, like the bad faith condition I will describe in the following chapter, externalizes responsibility and freedom, and thus brings about a decrease in human dignity.[82]

80 Benn, 1984, p. 229.

81 Fried, 1984, p. 218.

82 Moor (1997) adds another justification of privacy, beyond the one in terms of control/restricted account I have adopted here: privacy is also justified as an expression of the

Contrary to Kant, who believed that "there is no need of science or philosophy for knowing what man has to do in order to be honest and good," I contend that ethics and decision making should *always* be accompanied by knowledge related to the situation at hand. If we want knowledge to be considered a duty, we must commit ourselves to generating, distributing, and using knowledge in the service of personal, economic, and social development. Knowledge-as-duty is an ethical obligation not only to human beings living today but also to those of the future.

The vital importance of knowledge means that we must use great care in its management and distribution, and, as we have seen, there are several transdisciplinary issues related to this challenge: from promoting creative, model-based, and manipulative thinking in scientific and ethical reasoning to deepening the study of "epistemic" and "moral mediators"; from the interplay between unexpressed and super-expressed knowledge and their role in information technology to the shortcomings of the existing mechanisms for promoting rational and ethical argumentation. A lack of appropriate knowledge creates a negative bias in concrete moral deliberations and is an obstacle to responsible behavior, which is why I have emphasized the potential benefits of new "knowledge communities," of implementing ethical "tools" like ESEM capacities and trading zones, and of acknowledging the status of humans beings as "knowledge carriers."

We must also remember, however, that the proliferation of information carries with it some risks, primarily in the areas of identity and cyberprivacy. The internet and various databases contain an astounding volume of data about people, creating for every individual a sort of external "data shadow" that identifies or potentially identifies him or her, and the possible consequences of this must be examined. To avoid being ostracized or stigmatized, people must be protected not only from being seen but also from "feeling" too visible, which impairs the ability to make choices freely. And when sensitive information about a person is properly shielded, he or she is much less susceptible to exploitation and oppression by those seeking to use data unscrupulously.

These are some of the external conditions required for us to improve human dignity, but what of internal conditions? That is, how can we orient our thinking so that we are in the best position to embrace the knowledge-as-duty concept? Just as we must build greater knowledge about technology, so too must we strive for deeper understanding of ourselves. I believe that increasing awareness of ourselves, comprehending how and why we

core values that all normal humans and culture need to survive: life, happiness, freedom, knowledge, ability, resources, and security. In this sense, privacy is justified as an essential part of the central framework of values for a computerized society.

think, what we think, is another critical component of human dignity. Before we can begin to reap the benefits of value that others give to us, we must assess whether we have ascribed appropriate value and agency to *ourselves*, which brings us to the focus of chapter 5: the link between freedom and responsibility and their relationship to some internal conditions involving bad faith and self-deception.

5

Freedom and Responsibility

Bad Faith

> So that in the nature of man, we find three principal causes of quarrel. First, competition; secondly, diffidence; thirdly, glory.
>
> Thomas Hobbes, *Leviathan*

As we have seen, it is critically important that we explore the ethical implications of technology so that we understand its relationship to, say, the medicalization of life and cybernetic globalization. While these are crucial issues, they are limited to the collective level of awareness and fail to address other factors. What about the level of personal consciousness? How does what we know as individuals affect human dignity? If we lack awareness of how we think, of how and why we construct our beliefs, will we be equipped to make the best possible decisions? How does the manner in which we obtain and manage information shape our identities and self-perceptions? Our personal relationship to knowledge plays a part in determining our own level of freedom as well as our capacity for improving the lives of others. If we fail to interrogate such issues, we deprive ourselves of a way to improve the lives of other people, and consequently, knowledge as duty must be embraced not only nationally or globally but also at the local level, individual by individual.

I believe some answers to these questions can be found in Jean-Paul Sartre's notion of bad faith, which, put simply, is a kind of falsehood that involves lying to oneself. In the human condition of bad faith, people treat themselves as means; they ignore or jettison the concept of choice in some respect because it is somehow vexing or burdensome, and in doing so they relinquish freedom and externalize responsibility. When in bad faith, we do not "respect" ourselves as highly commodified means/things, and so we do not recognize that instrumental condition as being an end in itself.

Using Sartre's ideas as a starting point, I hope to show that a great deal of human suffering due to bad faith results from the woefully inadequate and "incomplete" kinds of knowledge and information currently available to us. The discussion will build on ideas introduced at the beginning of this book: it seems that it is because people do not know how to "respect people as things" (that is, to respect people as facticity and being-for-others) that they do not respect or appreciate themselves and, consequently, put themselves at greater risk of falling into the trap of bad faith. From the perspective of bad faith, if we do not "respect human beings as things," then we deny them various values – economical, cultural, moral, and so on – that we blithely attribute to external objects, external situations, and roles.

Once we fall into the condition of bad faith, we suffer a terrible consequence: bad faith denies us a full range of choice and erodes our ability to direct our own destinies and, consequently, strips away a measure of our freedom. Worse still, this interplay between bad faith and decreased freedom and responsibility gnaws away at human dignity and leaves us more vulnerable to exploitation and other forms of ill treatment. The pop psychologists on radio and television are fond of saying that the first step toward healing is to admit that there is a problem, and there may be some truth to that facile cliché: acknowledging and understanding our "condition" can weaken bad faith's grip and be extraordinarily helpful in boosting freedom, securing ownership of our own destinies, and establishing incentives to adopt more responsibility. Accruing knowledge, then, is also a duty in the context of personal identity; it can help us to manage our beliefs and guide us individually toward collective solutions to a number of difficult problems. I offer two such examples later in this chapter as I examine the increasing commodification of various aspects of our humanity and the paradoxical role that "respecting people as things" plays in modern warfare.

BAD FAITH

Denying the fact that choices exist is one way for human beings to avoid anxiety: by creating a simple world in which there are no alternatives, people shield themselves from the responsibility of decision making. This self-deception, or *bad faith*, creates a situation in which human beings relinquish freedom and externalize responsibility.[1] Because bad faith involves lying to oneself rather than to others, it is a particular kind of falsehood that directly affects the "constitution" and the "inner structure" of our consciousness, and consequently it must be distinguished from simple lying in general.[2]

[1] Van den Hoven, 2000.
[2] Sartre, 1956, p. 48.

Some of Jean Paul Sartre's seminal considerations on bad faith can help to elucidate the idea:

Let us consider this waiter in the café. His movement is quick and forward, a little too precise, a little too rapid. He comes toward the patrons with a step a little too quick. He bends forward a little too eagerly; his voice, his eyes express an interest a little too solicitous for the order of the customer. Finally there he returns, trying to imitate in his walk the inflexible stiffness of some kind of automaton while carrying his tray with the recklessness of a tight-rope walker by putting it in a perpetually unstable, perpetually broken equilibrium which he perpetually reestablishes by a light movement of the arm and hand. All his behavior seems to us a game. He applies himself to chaining his movements as if they were mechanisms, the one regulating the other; his gestures and even his voice seem to be mechanisms; he gives himself the quickness and pitiless rapidity of things. He is playing, he is amusing himself. But what is he playing? We need to watch long before we can explain it: he is playing at *being* a waiter in a café.[3]

Sartre continues by saying that the waiter is realizing his "ceremony," much like a grocer who is required by society to limit himself to the functions of that role: "a grocer who dreams is offensive to the buyer, because such a grocer is not wholly a grocer."[4] Based on the description of the waiter, one surely cannot deduce that he does not generate any kind of reflective "judgments or concepts concerning his condition"[5]:

He knows well what it "means": the obligation of getting up at five o'clock, of sweeping the floor of the shop before the restaurant opens, of starting the coffee pot going, etc. "He knows the rights which it allows: the right to the tips, the right to belong to a union, etc. But all these concepts, all these judgments refer to the transcendent. It is a matter of abstract possibility, of rights and duties conferred on a "person possessing rights." And it is precisely this person *who I have to be* (if I am the waiter in question) and who I am not. I do not wish to be this person or that, I want this person to be different. But rather there is no common measure between his being and mine. It is a "representation" for others and for myself, which means that I can be only in *representation*. But if I represent myself as him, I am not he; ... I can only play *at being* him; that is imagine to myself that I am he.[6]

Freedom, Responsibility, and Bad Faith

As we will see, bad faith affects freedom and responsibility: what the fellow in the previous example is trying to realize, Sartre says, is the

[3] Ibid., p. 59.
[4] Ibid.
[5] Cf. the section later in this chapter "Owning Our Own Destinies."
[6] Sartre, 1956, pp. 59–60.

"being-in-itself"[7] of the waiter, as if he were unable to confer the appropriate value and urgency to his position or as if it were not his free choice whether "to get up each morning at five o'clock or to remain in bed, even though it meant being fired."[8] To summarize in a few words: the man is a waiter when he is in the *mode of being what he is not*. So he is thinking "Okay, I'm a man, but when I work as a waiter, I become not what I am (a man) but what I'm not (a waiter)." While in the waiter mode, he is ignoring/denying certain possibilities – that he could choose *not* to behave as a waiter, for example – and his freedom is, consequently, curbed by his self-imposed restriction. He is in a condition of bad faith, deceiving himself by constructing a limited reality that does not take into account the full range of choices available to him, and this, alas, is a condition in which many people live all their lives. It is from himself that he is hiding the truth; the deceiver and the deceived coalesce into a single consciousness in a way that must be distinguished from true mental illness or malfunction of consciousness, as discussed in chapter 3 (in the section "Is Conscious Will an Illusion?").

According to psychoanalytic tradition, bad faith can be understood as part of the dynamics of the unconscious. If we hypothesize a Freudian "censor," we can reestablish the duality of the deceiver and the deceived: the psychic whole, the "single consciousness" mentioned earlier, is then cut into two parts, the "Id" and the "Ego." In the case of psychoanalysis, the notion of bad faith manifests itself as the idea, so to say, of a lie without a liar,[9] a notion that seems to relieve the deceiver of agency and choice. In this way of thinking, a person cannot accept responsibility for practicing bad faith because of the opacity of the trigger for such behavior. But Sartre maintains – and I agree with him – that his idea of bad faith is not undermined by the fact that this self-deceit occurs at an unconscious level; being in bad faith does not mean we are *aware* that we are living in the condition. After all, "there exists an infinity of types of behavior in bad faith which explicitly reject this kind of explanation because their essence implies that they can appear only in the translucency of consciousness. We find that the problem which we had attempted to solve is still untouched."[10]

7 It is difficult to give abstract definitions of Sartrian "being-in-itself" and "being-for-others." Their meanings will be clearly explained in the context of the following discussion. The "being-for-others" is perceived by people "over there," "out in the world," as a waiter, for example; the "being-for-itself," on the other hand, is the being as perceived by itself.

8 Sartre, 1956, pp. 59–60.

9 Ibid., p. 51.

10 Ibid., p. 54.

Consciousness, in its total "translucency," affects *itself* with bad faith, as Sartre points out, and thus it would seem that a being in bad faith must be conscious of its state, "since the being of consciousness is consciousness of being."[11] In reality, however, Sartre says the entire psychic system is, in this case, "annihilated": "We must agree in fact that if I deliberately and cynically attempt to lie to myself, I fail completely in this undertaking: the lie falls back and collapses beneath my look: it is ruined from behind by the very consciousness of lying to myself which pitilessly constitutes itself well within my project as its very condition."[12] In other words, for self-deceit to be successful, the psychic whole must be maintained and internal dialogue must be quashed; only then can a kind of self-protection be achieved.[13]

There are countless examples of this kind of performance: the violinist who plays at being a violinist, the plumber who plays at being a plumber, the professor who plays at being a professor, all of whom are simultaneously different, "transcendent" people, but of course they are not consciously aware of such dichotomies. Such a situation establishes a dichotomy between the "being-for-others" and the "being-for-itself." The being-for-others is perceived by people "over there," "out in the world," as an intellectual or waiter, for example; the being-for-itself, on the other hand, is the being as perceived by itself. In the case of the waiter, bad faith arises from the multiplicity inherent in human reality, realized here by the dichotomy between the "being-for-others" and the "being-for-itself."

Another interesting way to understand the multifaceted nature of this reality is to explore a second dichotomy, that of "transcendence/facticity," as Sartre says.[14] Consider, for example, Sartre's scenario involving a man who expresses interest in a woman and her subsequent internal response:

> If he says to her "I find you so attractive!" she disarms this phrase of its sexual background; she attaches to the conversation and to the behavior of the speaker, the immediate meanings, which she imagines as objective qualities. ... She is profoundly aware of the desire which she inspires, but the desire cruel and naked would humiliate and horrify her. In order to satisfy her, there must be a feeling which is addressed wholly to her *personality* – i.e., to her full freedom – and that would be a full recognition of her freedom. But at the same time this feeling must be wholly desire; that is, it must address itself to her body as object. This time then she refuses to apprehend the desire for what it is; she does not even give it a

[11] Ibid., p. 49.

[12] Ibid., pp. 49–50.

[13] Some authors consider the notion of bad faith implausible (Paluch, 1967; Haight, 1980). Others think that bad faith involves a division in the self (Davidson, 1985; Pears, 1986; Rorty, 1988, 1996), an explicit intention (Davidson, 1985; Pears, 1986; Rorty, 1988; Talbott, 1995), a nonintentional process (Johnston, 1988; McLaughlin, 1996; Lazar, 1999; Mele, 2001), or condemnation/remorse (Jankélévitch, 1966).

[14] Again, it is difficult to give an abstract definition of the Sartrian dicothomy "transcendence/ facticity," but its meaning will be clearly explained in the context of the following discussion.

name; she recognizes it only to the extent that it transcends itself toward admiration, esteem, respect and that it is wholly absorbed in the more refined forms which it produces, to the extent of no longer figuring anymore as a sort of warmth and density. But then suppose he takes her hand. This act of her companion risks changing the situation by calling for an immediate decision. ... The young woman leaves her hand there, but she *does not* notice that she is leaving it. She does not notice because it happens by chance that she is at this moment all intellect. ... The hand rests inert between the warm hands of her companion – neither consenting nor resisting – a thing.[15]

The woman uses various attitudes to maintain her state of bad faith and, one assumes, to protect her view of herself as "virtuous." She ends by considering herself as "*not being* her own body" and contemplates it as a passive object to which "events can happen but which can neither provoke them nor avoid them because all its possibilities are outside of it."[16]

In this case, it is clear how bad faith is enacted by generating "contradictory concepts which unite in themselves both an idea and the opposite of that idea." The woman takes advantage of the dual properties of human beings, the human condition of being at once a *facticity* and a *transcendence*: "bad faith does not wish either to coordinate them nor to surmount them in a synthesis. Bad faith seeks to affirm their identity while preserving their differences. It must affirm facticity as being transcendence and transcendence as being facticity, in such a way that at the instant when a person apprehends the one, he can find himself abruptly faced with the other."[17] In this way, the woman of Sartre's example can accomplish her goal of attaining "transcendent" sexual status. We are always juggling multiple roles, many of which serve more than mere social functions. As Sartre notes, "I am never any one of my attitudes, any one of my actions. ... The good pupil who wishes to *be* attentive, his eyes riveted on the teacher, his ears open wide, so exhausts himself in playing the attentive role that he ends up by no longer hearing anything."[18]

Later, in the section "Owning Our Own Destinies," I will argue that while bad faith is indeed contradictory, contradiction is not its central, defining feature. Bad faith is contradictory from a cognitive point of view because when I am in bad faith, "I also am what I am not." But human cognitive behavior manifests other contradictions that do not involve bad faith. Like Sartre, I consider bad faith in social contexts to be a kind of spiritual and moral failure, for reasons I will explain in the following section. But bad faith also has its benefits: for example, it could be

[15] Sartre, 1956, pp. 55–56.
[16] Ibid., p. 56.
[17] Ibid.
[18] Ibid.

considered a plausible way to maintain a certain degree of happiness, for as the old saying goes, "Ignorance is bliss."

Also, overcoming bad faith certainly requires that we eliminate the contradiction on which it depends, but only because it is the "bad faith contradiction": indeed, I maintain that contradictions in other types of behavior do not result in the same kind of moral damage – the loss of freedom and responsibility, for example – wrought by bad faith (see the section, "Owning Our Own Destiny"). What is the role of information and knowledge in bad faith? Must I actually "know" the truth on some level in order to conceal it from myself?

FREEDOM, INCOMPLETE KNOWLEDGE, AND THE EXTERNALIZATION OF RESPONSIBILITY

For people in bad faith, the "being-for-others" is an external instrumental condition that does not fully encompass their own identity. As the traditional distinctions between humans and things grow more indeterminate in our technological era, our status as hybrid people reinforces this "being-for-others," and this fact paired with the increasing "commodification" of ourselves gives rise to a great many more occasions for bad faith. In bad faith, we do not "respect" ourselves as means/things, and so we do not recognize that instrumental condition as involving an end in itself.

As we have illustrated in the previous section, when we are in bad faith, we are always dealing with more than merely social positions. Sartre writes, "I am never any one of my attitudes, any one of my actions," and this statement seems to offer the perfect way to avoid responsibility. To follow his line of thinking, if I simply am what I am, if I have just one stable, clearly defined role, then I can see the truth in criticism offered by others and will be compelled to accept it. But thanks to the fact that *I am also what I am not*, I am not subject to any criticism: "I do not have to discuss the justice of the reproach. As Suzanne says to Figaro, 'To prove that I am right would be to recognize that I can be wrong'. I am on a plane where no reproach can touch me since what I really am is my transcendence," as is the case with Sartre's woman, who considers herself pure transcendence so that she can disavow the horrifying power of her body and the unspeakable urge it produces – desire.[19]

Let us come back to the idea of bad faith as a condition in which we abnegate our own freedom and responsibility because they are too burdensome, as in the case of the waiter or the coquette. In the case of the waiter, equal dignity arises from being-for-others (our Kantian "people as means") and being-for-itself (our Kantian "people as ends"), but this

[19] Ibid., p. 57.

dignity is inhibited by a continuous "disintegrating synthesis" and a "perpetual game of escape from the for-itself to the for-others and vice versa." A similar condition occurs in the dichotomy of transcendence/facticity (or ends/means): here the disintegrating role played by the woman as a "passive object among other objects," as a "means/body," is even more evident, Sartre appropriately says, and she becomes a "thing."[20]

The woman, then, avoids distress by casting herself as innocent of desire, as neither fueling it in her companion nor experiencing it herself. This contradictory interplay of bad faith externalizes responsibility and, consequently, negates any anguish that might result from a particular choice. I hypothesize, however, that if this phenomenon arises from an inability to tolerate anxiety, then bad faith is, in fact, the result of a lack of (or ignorance of) information and knowledge.

The cognitive and emotional aspects of bad faith are deeply intertwined with one another.[21] The waiter assigns an emotional attitude to being a "transcendent" waiter that contrasts with the facticity of the social role; similarly, the woman assigns an emotional attitude to being "transcendent" that contrasts with the reality of her body: in both cases, the emotional attitude generates stress, which then triggers the need to reduce it by activating a (false) sense of control. The anxiety generated in bad faith is connected to people's attempts to avoid or to approach particular subjective goals, and it is certainly linked to the emotional need for an acceptable degree of happiness, as I have already pointed out.[22] Because of this role of emotions, we can affirm that bad faith cannot be considered completely intentional: in this sense emotions (unlike feelings)[23] occur unconsciously.

This ignorance, which is at the root of all bad faith, leads us to regard external things in their myriad forms – mediators, institutions, social relationships, tools, bodies, and so on – as distant and oppressive, a burden that we, who are so weak, cannot shoulder. Anxiety is the inevitable result of such thinking! A lack of knowledge may initially seem to offer protection, and many people prefer to play the fictitious bad faith role rather than increase their understanding of a situation. It is such ignorance, however, that in turn breeds human anguish: when we choose to avoid new knowledge, we activate bad faith and become "thing-like" objects that are bound by the uncontrollable constraints of an external automatism.

[20] Ibid., p. 58.
[21] On the central role of the emotions in bad faith, cf. Sahdra and Thagard, 2003; de Sousa, 1988; and Lazar, 1999.
[22] Cf. Sadhra and Thagard, 2003, p. 222.
[23] Damasio, 1999.

For instance, in Sartre's examples, the lack of knowledge usually involves issues of "facticity" and "being-for-others" that are inherent in bad faith devices: the facticity of the "being-for-others" immediately implements a "fictitious" role. Because people do not know how to "respect people as things" (that is, how to respect people as facticity and being-for-others), they do not respect and appreciate themselves, which springs the trap of bad faith. In the case of the woman, denying or failing to understand the nature of the biological body and its sexual and instinctive needs creates fertile ground for the contradictory bad faith interplay. The same conditions occur in the case of a being-for-others who is not granted the dignity that is appropriate to his or her particular social role and, as a result, never attains the positive, objective value of a "thing" that exists "over there" in the world. The same dignity that is not attributed "over there," in the external world, to the "being-for-others" is in turn not attributed to oneself as "being-for-itself," so that one plays only "the external part." Hence human beings activate bad faith and are reduced to reified and partial "thing-like" objects.

In summary, in terms of bad faith, when we do not acknowledge the full range of human worth – economic, cultural, moral, and so on – we fail to respect people as things and end up attributing these important values only to external objects, situations, and roles. As a result, when these values are not appropriately reassigned to ourselves, when we fail to translate them into human terms, people are denied the human dignity they deserve. Sometimes values of this kind exist, but they have been damaged or diminished; in the worst cases, however, such values were never present to begin with. These situations result from a lack of know-how – general, scientific, legal, economical, and moral – about objects (things, external roles and relations, etc.) and their value (sometimes intrinsic, as in the biological needs of the human body). The "being-for-others" is constructed by people in bad faith as a safe external entity that cannot participate in their own identity. To reintroduce the Kantian terms we used before, when in bad faith, we "treat" a person as means (a "being-for-others," in the case of the waiter) and accord him or her only the negative aspects of "thing-hood." If we fail to associate external things' positive qualities with the person, if we do not "respect" the human being as a means/thing, then we do not recognize that instrumental condition as incorporating ends in itself.

OWNING OUR OWN DESTINIES

For human beings, the condition of bad faith clearly represents diminished freedom and choice: in "playing" a role, we passively obey external limitations. An example could be the Sartrian waiter who does not choose to defend his rights as a worker: yes, he is a waiter, but in reality he is not, so the external role he plays is merely a limiting horizon. When we decide to "make do" with or accept a situation in this way, the ownership of our own destinies is assaulted, just as it is when we face the problems of privacy discussed in the chapter 4 section "Identity and Privacy in the Cyberage." A lack of privacy could also eliminate protection against insults that "slight the individual's ownership of himself," as happens with, say, prisoners or slaves: as a consequence, people will say *mine* with less authority and *yours* with less respect. Something similar occurs in bad faith, where the autonomy of the "being-for-itself" is sacrificed to the automatism of the "being-for-others." Or imagine the self-censorship and diminished range of choices suffered by a person who is known to be afflicted by a serious disease – in this case, too much information about him or her is externalized.

In this instance, too much of our identity and data is externalized in distant objects not sufficiently dignified as "things." Hence, the possibility of being reified in those externalized data increases, while our dignity decreases. This loss of freedom is, in turn, immediately externalized and therefore becomes apparent to *others*, so that we are automatically deceived by our duplicity and, consequently, less respected. The result is clear: our own and others' moral dignity is affected and diminished. When we lose ownership of our actions for whatever reason – bad faith, in this case – we lose responsibility for them, given the fact that morality is related to the ownership of our destinies and to the right to choose in our own way. Sartre clearly says that in spite of the fact that bad faith is an unstable phenomenon, it can feel like a very normal way of life, a kind of "being in the world, like walking or dreaming." Bad faith is not at all like the simple, compartmentalized "cynical lie": it configures all aspects of our lives.[24]

Like Sartre's waiter and coquette, the character Dimmesdale in Nathaniel Hawthorne's *The Scarlet Letter* (1850)[25] engages in self-deception that generates tension between transcendence and facticity – that is, between the "being-for-itself" and the "being-for-others." As he righteously and publicly condemns Hester Prynne for bearing a child out of wedlock, a child that he in fact fathered, he enters the condition of bad faith, denies responsibility, and vitiates his own freedom.

It is not unusual to connect the idea of bad faith with the Marxian concept of false consciousness,[26] and it seems plausible – at least in the social environments of rich countries – to think that limited freedom generated by bad faith can facilitate external oppression and increase one's vulnerability to subjection by another individual or group. When

[24] Ibid., p. 68 and p. 67.

[25] Sadhra and Thagard (2003) magnificently discuss this case of bad faith.

[26] Coombes, 2002.

those in bad faith externalize their own freedom and responsibility in order to avoid emotional discomfort, they can more easily be exploited and kept in a position of subordination. Being in a condition of bad faith may mean accepting as unavoidable external values that run counter to one's own interests, which then perpetuates bad faith: individual action and free choice become more difficult, which, in turn, renders the ownership of one's own destiny less likely. The Marxist tradition uses the word "alienation" to express the state of people who lack full personal control of their choices and, thus, of their futures.

Increasing the personal freedom of those in bad faith requires a special effort, and the potential for success varies greatly depending on the extent of external constraints and the strength of internal will in each situation. It is difficult to manage all the images and constructions of ourselves as we attempt to increase personal freedom; our self-images are often entangled with the images others have of us, and it is difficult to tease out the difference between the two. Even when they are contradictory, we may not necessarily perceive them as such.

When viewing the world from the vantage point of bad faith, we see ourselves in ways that may seem to be in harmony but are actually in conflict, and we see others and external things with even less clarity and consistency. How many times do we consider people as a means one moment and as an end the next, without ever being aware of any contradiction in our thinking? Even if we are aware of our conflicting views, we may often feel powerless to reconcile the two contradictory perceptions.

Let me offer a Kantian example of such a conflict from the world of academia. Say a professor collaborates with a student on research and publication and, in doing so, considers him not solely as a "means" but also as a respectable "end," as Kant advocates. Nevertheless, this same professor neglects the other needs of the Ph.D. candidate, does not help to establish his career, and avoids all opportunities to help, so that the student, who trusted that the professor would assist him, is deceived. The student, then, is treated not solely as a means but also as an end (the end of performing and publishing research), but when his other needs (his employment needs after having earned his Ph.D., for example) are not addressed, his image is dichotomized and his dignity is therefore not respected. What is ironic is that the professor does not perceive this gap and the contradiction it creates because he regards the student as a thing-person;[27] the professor does not care because he has not endowed this student with the dignity befitting someone who is potentially ready to become a professor himself. The student, in turn, thinks (in bad faith)

[27] Cf. chapter 1, of this book, the section "The Dispossessed Person, the Means-Person, and the Thing-Person."

of himself in the same way as he "plays" the role of a researcher/potential teacher. When the student gives up his freedom and responsibility, the professor can easily exploit him, and the bad faith circle is completed in a kind of Marxian perfection.

As they unconsciously deceive those around them, people in bad faith cause damage by weakening others' ownership of their own destinies. People in bad faith generate images of themselves in others that reflect a condition very much like the one of bad faith, one that consists of a gap between beliefs and reality. It is clear that if we deceive ourselves and others through the bad faith effect, it becomes very difficult to treat anyone – others or ourselves – with the appropriate level of respect; it is almost impossible to evaluate accurately the merits and demerits of people who deceive themselves or who are deceived. The consequence is obvious: the ownership of our own destinies is further weakened, and the shining "kingdom of ends" that Immanuel Kant suggests we "bring into existence ... by our conduct" becomes even more remote.[28]

The greatest obstacle to owning our destinies – a task that is primarily ethical – is a lack (or ignorance/repression) of knowledge. As I explained in the previous chapter, today's collectivities involve a huge flow of information, and as a result, adequate knowledge is becoming essential for effective functioning in most modern settings. Indeed, images of ourselves, of others, and of external things are multiple, unstable, and intertwined and, at the same time, fragmented and murky – all of which creates the conditions necessary for bad faith. I maintain that understanding the dynamics of bad faith is our best weapon against it, and that such knowledge is extraordinarily helpful in bolstering freedom, ensuring our ownership of our own destinies, and establishing incentives to adopt more responsibility. This notion brings us back to the motto "knowledge as duty" from chapter 4: knowledge, when available, exploitable, and appropriate, is a critical factor in the development of freedom and responsibility.

There is in the literature much research on self-deception and weakness of will, also called *akrasia*.[29] Contrary to my point of view is Amelie Rorty's contention[30] that self-deception jeopardizes rationality's status as the most effective of adaptive mechanisms. She describes what she sees as the benefits of thinking that only oneself is capable of constructing "second-order policies" and beliefs, and that when these conflict, they can immediately be reconciled by adaptivity and rationality: "The animus that

[28] Kant, 1964, p. 104.

[29] Cf. also footnote 13. Other philosophical and psychological perspectives on bad faith and weakness of will are illustrated in Davidson, 1970; de Sousa, 1970; Rorty, 1972; and Schiffer, 1976.

[30] Rorty, 1972.

guides me in this matter is the conviction that loose and indiscriminate reading of Kant has led us to overemphasize the unity of persons. While there were very good reasons for doing this, we need not be overwhelmed by these reasons. I think we can account for ourselves as believers and agents without presupposing the Unity or the Strict Identity of the Ego."[31]

My response is: exactly! Amelie Rorty is correct, however, when she questions whether the unity of consciousness legitimates irresponsibility. Bad faith does not result from some sad, pathological conflict between the self and the unconscious; rather, it occurs, as Sartre says, in the "translucency" of the consciousness. It is at this level that, as I will illustrate later, knowledge and responsibility lead us to adopt *one or more* roles we play.

Contrary to Donald Davidson,[32] John Searle[33] holds that weakness of will does not mean that agents act contrary to their intentions, because they did not really have "unconditional" intentions to perform certain actions. Instead, Searle argues that the unavoidable gap between the intention-in-action and the action itself always creates an opportunity for weakness of will. People's motivations may change; they may adopt different reasons to act, or they may acquire new desires: "It is not the case that everything that one judges to be best to do, one really wants to do, and it is not the case that when you have made up your mind and you really want to do something, that you therefore necessarily do it."[34] At any moment after the first intention, the range of possible choices is so broad that at least some of them are likely to be attractive to the agent.

Searle considers self-deception a kind of conflict avoidance achieved by suppressing unpleasant ideas that threaten what is desired. In this case, if one is to achieve one's desire, such threats must be suppressed. He thinks that self-deception "logically requires the notion of unconscious"[35] and would take the following form: "The agent has the conscious state: I believe not *p*. He has the unconscious states: I have overwhelming evidence that *p* and want very much to believe that not *p*."[36] For example, a weak-willed person might say to himself;

> "Yes I know I shouldn't be smoking another cigarette and I have made a firm resolve to stop, but all the same I do want one very much; and so, against my better judgment, I am going to have one." But the self-deceiver cannot say to himself, "Yes I know that the proposition I believe is certainly false, but I want

[31] Ibid., p. 401 and p. 405.
[32] Davidson, 1970.
[33] Searle, 2001.
[34] Ibid., p. 228.
[35] Ibid., p. 235.
[36] Ibid., p. 236.

very much to believe it; and so, against my better judgment and knowledge, I am going to go on believing it."[37]

Contrary to Searle, Sartre emphasizes that the analogous case of bad faith occurs in the "translucency of consciousness," where the agent lies to him- or herself and does not recognize inconsistency and irrationality, which, therefore, do not need suppression. I, however, agree with Sartre: while some may argue that weakness of will arises from an unconsciously generated and therefore unavoidable gap between the intention-in-action and the action itself, I believe we can interpret Sartre's treatment of bad faith as an affirmation that knowledge is vitally important to freedom and dignity, as I have illustrated. I think that the problem posed by Searle can be solved when we consider that, as I have previously illustrated, it is the presence of a profound emotional appraisal in the condition of bad faith that accounts for its unconscious side.

Finally, Daniel Dennett believes that bad faith, in the form of what he calls "local fatalism," is the exception rather than the rule: he writes that "it is our good fortune that these conditions are abnormal in our world."[38] Like Sartre and Pirandello, however, I think that bad faith is, unfortunately, widespread and lurks unnoticed in many people's lives.

"ONE, NO ONE, AND ONE HUNDRED THOUSAND"

In his 1926 novel *One, No One, and One Hundred Thousand*,[39] the Italian writer Luigi Pirandello describes a man, a small town squire, who looks in the mirror one day, touches a nostril, and feels some pain. His wife tells him his nose tilts to the right, something he had not realized before. Reflecting upon this sudden revelation, he concludes that he possesses different personalities. Is not this the objective minimal condition of bad faith? So he begins a search to check his various selves. After a series of strange incidents, he is deserted by his wife and declared insane. The court gives his money to a poorhouse; he becomes its first guest. He becomes the "no one" of the book's title, and, by being no one, the gentleman becomes everyone. He can be reborn again and again. "I am I and you are you," declares the squire, speaking as the first person narrator of the novel. Immediately after that statement, he says: " . . . a minute ago, before you were in this situation, you were another; and that's not all of it, you were also a hundred others, a hundred thousand others. And nothing can be done about it, believe me, there's nothing to be

[37] Ibid., p. 234.
[38] Dennett, 1984, p. 106.
[39] Pirandello, 1990.

amazed at. See, on the contrary, if you can be so sure that by tomorrow you will be what you assume you are today."[40]

The gentleman is also clearly aware that human beings are always morally "building" not only their own selves, but also the selves of "others" and, so to say, the selves of "things":

> Do you believe you can know yourselves if you don't somehow construct yourselves? Or that I can know you if I don't construct you in my way? ... We can know only what we succeed in giving form to. But what kind of knowledge can that be? Is this form perhaps the thing itself? Yes, both for me and for you, but not the same for me as for you: in fact, I don't recognize yourself in the one I give you; and the same thing is not identical for all and for each of us it can continually change, and indeed, it does change continually. Still, there is no other reality outside of this, the momentary form we manage to give to ourselves, to others, to things.[41]

Hence, the selves that people can construct are many. Pirandello's character goes on:

> So let's skip the name, and we'll also skip the features – now that I am in front of the mirror I clearly realized how I was necessarily unable to give myself an image different from the one with which I conceived myself – I felt also these features alien to my will and spitefully contrary to any desire that might arise in me to have others, different from these, that is this hair, of this color, these eyes as they are, greenish, and this nose and this mouth. We'll skip also my features, as I said, because, after all, I had to accept the fact that they could even have been monstrous and I would have had to keep them and resign myself to them, if I wanted to live. But they weren't, and so, all in all, I could even be satisfied with them.[42]

In the end, Pirandello's protagonist says: "I no longer look at myself in the mirror, and it never even occurs to me to want to know what has happened to my face and to my whole appearance. The one I had for the others must have seemed greatly changed and in a very comical way, judging by the wonder and the laughter that greeted me."[43]

Another novelist, Italo Svevo, also explores the multiple facets of our selves in his 1923 novel *Confessions of Zeno*.[44] The pliant protagonist of the story is, among many other things, a businessman, a guilt-ridden adulterer, and a nicotine addict, and the dense novel unfolds through the narrator's self-revelations, which he undertakes as part of his psychoanalytic treatment. Zeno will never find a cure for afflictions that seem

[40] Ibid., p. 33.
[41] Ibid., p. 41.
[42] Ibid., p. 51.
[43] Ibid., p. 159.
[44] Svevo, 1958.

radical and indispensable. His reflections are complicated rationalizations that take place at a conscious level, very similar to the ones involved in bad faith. He formulates numerous reasons why his "last cigarette" need not truly be his last; he continuously tries to convince himself that he loves his wife; he tirelessly justifies an awkward affair, all the while exhibiting a passive, submissive mindset in which his will to act seems paralyzed. "My resolutions are less drastic and, as I grow older, I become more indulgent to my weaknesses," Zeno proclaims early on. Later he is even more revealing when he admits that his "resolutions existed for their own sake and had no practical results whatever."[45] He acknowledges neither the freedom he has to make decisions nor the responsibility he bears for the choices he does make; instead, he describes his disappointing failures as inevitable. Bad faith has permeated Zeno's life, just as it does the lives of countless others, distorting his view of everything around him. It becomes a way of life in which fictions, dark motivations, subterfuges, feints, and games dominate human interaction and, as a result, drain the freedom from such transactions.

FREEDOM, RESPONSIBILITY, AND KNOWLEDGE

The characters of Pirandello and Svevo are excellent examples of men encumbered by an existential burden, a burden that in these cases is caused by the ineluctability of living with so many facets of the self. Anguish is so heavy that we prefer to externalize responsibility and reduce our freedom by playing roles in an automatic way. Limiting ourselves to these many roles may create an illusion of comfort, but, at the same time, doing so also curtails our own freedom as well as the freedom of others. As I have already illustrated, if we externalize responsibility by conducting ourselves as objects (cf. facticity and the "being-for-others" of Sartre's examples), we can fall into exploiting others who are deceived by our many facets or, conversely, create opportunities for them to engage in exploitation themselves: when we act in bad faith, (1) we may force other people to assume more responsibility in order to compensate for what we have neglected, or (2) we may be giving them greater opportunity to exploit us as we stumble around in a haze of self-deception.

For someone who is in bad faith, neither others nor other things are "really" included in his or her "self." To recall Sartre's example, we regard our jobs as external and abstract and therefore not really ours. The same occurs when we think of our bodies as something external to and unaffected by us, as in the case of the coquette. When in bad faith, we think of many things – jobs and bodies are just two examples – in

[45] Ibid., p. 9.

conflicting ways: as external things that both belong to us, and, at the same time, do *not* belong to us.

As Pirandello reminds us with his hundred thousand ways of thinking of a human being, when we consider other people, we spontaneously objectify them in many ways, including as things or even as tools. This tendency creates problems, of course, because people are unable to recognize themselves in these objectifications. I maintain that these gaps (between the "being-for-itself" and the "being-for-others") result not only from inadequate and "incomplete" knowledge and information, but also from inadequate emotional appraisals. Because we simply do not possess sufficient information about our jobs, bodies, actions, and so on, we fall into the bad faith disposition. This lack (or ignorance/repression) of knowledge favors emotions like anguish that human beings cannot tolerate, and, consequently, they seek refuge and protection in fictitious roles. In the passage quoted earlier, Sartre says the waiter "knows well what it 'means': the obligation of getting up at five o'clock, of sweeping the floor of the shop before the restaurant opens, of starting the coffee pot going, etc. He knows the rights which it allows: the right to the tips, the right to belong to a union, etc. But all these concepts, all these judgments refer to the transcendent. It is a matter of abstract possibility, of rights and duties conferred on a 'person possessing rights'."[46] I contend, however, that the waiter of this example does not really "know" (or has repressed or does not *want* to know) the majority of these "concepts" and "judgments" or, at least, does not "know" and "recognize" them in the correct way. A deeper understanding of his various roles and how they relate to each other would provide an incentive to take responsibility for his actions, and, at the same time, give him a great deal more freedom. The same holds true for the coquette, who would gain much from knowing and acknowledging her body's needs and her sexual role.

Sartre's man does not recognize the dignity of playing the role of the waiter because he does not fully comprehend that role as a "thing" endowed with objective worth. Values that are attributed "over there," in the external world, to the "being-a-waiter-for-others," are not considered applicable to ourselves as "beings-for-themselves": our value as things is not thought of as part of our identity, and, as a result, we cannot see it as something that contributes to our personal dignity. Consequently, we "play" at being a waiter, a professor, a wife, and so on, as external parts of ourselves, as an actor might step into a role on the stage. We disregard or neglect our value as things because it doesn't seem relevant to the "being-for-itself." Such a valuation may be considered inappropriate – if we see "thing status" as having great value, we may believe we do not deserve it

[46] Sartre, 1956, p. 59.

because of our lowly position; conversely, if we view it as having little worth, then using it to shape identity is simply humiliating.

INCONSISTENCIES AND MORAL BEHAVIOR

Even if a lack (or ignorance/repression) of knowledge is responsible for the condition of bad faith, it does not necessarily follow that any increment of knowledge can fix contradictions and guarantee access to new levels of freedom. A simple case illustrates my point. Contradictions can be generated from newly acquired knowledge, as in the so-called Evening Star, Morning Star case.[47] The planet Venus was once thought to be two different stars – one habitable, one not – and the discovery that these stars were actually a single entity created a contradiction. Such tension also occurs when we feel both desire for and aversion to the same object: "A gentleman of two centuries ago, realizing suddenly that defending his honor would, in the circumstances in which he found himself, simply be murder, might give up either his practical commitment to his honor or his commitment to not committing murder, or he might adjust his understanding of those commitment in hard-to-anticipate ways."[48]

Moreover, I do not think we can increase our freedom simply by eradicating the inconsistency of the condition of bad faith. Coherence in moral behavior is not necessarily good in itself; it can mask deeply held conflicting notions the agent has not yet acknowledged. The failure to perceive inconsistency in oneself is typical of those in bad faith. What people see as consistency in belief and behavior can later be exposed as inconsistency when they find new knowledge and adopt new perspectives (cognitive and emotional). Hence, we must stress here that coherence is both knowledge- and emotion-dependent. In instances of bad faith, having the appropriate information about our behavior, concerns, rights, and duties is the only way to reveal and reconcile the hidden inner inconsistencies.

I do not think all inconsistencies are necessarily bad from a moral point of view. When we pay attention to the problems of another person, we can reasonably adopt a perspective that is inconsistent with how we might view ourselves in a similar situation. For example, I believe it is important to fulfill one's responsibilities, and I would feel I had been neglectful if I failed to prepare an important article needed to participate in a particularly boring faculty meeting. I would not, however, think the same of a good colleague who is committed to taking care of bureaucratic chores. It is true that I have a greater moral commitment to science and culture than I generally do to bureaucracy, but because I also want to

[47] O'Neill, 2001.

[48] Millgram, 2001a, p. 15.

respect others' preferences and attitudes, I would adopt a stance toward my colleague that introduces an inconsistency in my moral belief system: one must fulfill one's responsibilities – usually. Cultural relativism has taught us that inconsistencies are sometimes good: I can personally hold two inconsistent moral attitudes, and they can both be correct because of the need to respect cultural diversity among human beings.

Nevertheless, inconsistencies can be linked to bad faith, as is the case of people with incoherent intentions – people who, for example, commit themselves to an end but refuse to pursue the means to it. The refusal may result from a lack of knowledge or, possibly, from knowledge that is available but makes pursuing the end seem an overwhelmingly difficult task or an anguish-producing endeavor. As in the case of those who doggedly cling to an external appearance of coherence, the inconsistency is not perceived as such by the moral agent.

To conclude, coherence and incoherence must always be seen as interrelated features of situated concrete cases; moreover, coherence and incoherence are knowledge-sensitive, and surely there does not exist a kind of universal and "secured" Hegelian *Aufhebung* that, through overcoming inconsistencies, automatically generates good, freedom, and responsibility.[49]

PRIVACY AND BAD FAITH

The lack of privacy is another issue to consider in a discussion of bad faith. By safeguarding privacy, we protect people's ability to develop and realize projects in their own way. If respect for others is related to respect for their rights, privacy is one of the most important rights. As illustrated in the previous chapter, "monitored" individuals are disinclined to act naturally for fear of revealing negative aspects of themselves, and their resultant artificial behavior may lead others to see them as false and cold. Unnatural behavior of this sort can affect the attitude of those whose esteem, love, and friendship a person desires; it is a humiliating self-betrayal, a self-deception that shares a feature with bad faith: both conditions externalize responsibility and freedom. Of course, as we also discussed in chapter 4, privacy's veil of secrecy can facilitate dishonesty and wrongdoing – remember the example of a family that conceals a member's violence toward a child, a family's nepotistic solidarity, the abuse of women. But in the case of the watched individual, deception arises not from privacy, as in the case of family violence, but from the immoral deprivation of it. In the case of bad faith, deception results from

[49] In chapter 6, I will illustrate in detail the problem of inconsistencies in moral reasoning and knowledge. See the section "Comparing Alternatives and Governing Inconsistencies."

a situation that is not perceived as immoral either by the individual or by the majority of those around her.

From a more general philosophical point of view, being under the scrutiny of others helps us to become aware of ourselves as knowable "things" that are endowed with certain features. Being monitored by others challenges and "limits" our primitive consciousness of pure freedom as subjects and choosers; it creates a hostile relationship that fosters resentment:

> Ego is aware of Alter not only as a fact, an object in the world, but also as the subject of a quite independent world of Alter's own, wherein Ego himself is mere object. The relationship between the two is essentially hostile. Each, doubting his own freedom, is driven to assert the primacy of his own subjectivity. But the struggle for mastery, as Sartre readily admits, is a self-frustrating response; Alter's reassurance would be worthless to Ego unless it were freely given, yet the freedom to give it would at once refute it.[50]

Benn ethically reinterprets Sartre's philosophical concern in terms of privacy: it is not scrutiny itself that is always offensive, but rather "unlicensed" scrutiny. The latter, even if it does not cause damage, can be resented as an impertinence. In a sense, unlicensed scrutiny affects people's self-perception as actual or potential choosers, and thus influences their freedom and the possible adoption of responsibility: "To conceive someone as a person is to see him actually or potentially as a chooser, as one attempting to steer his own course through the world, adjusting his behavior as his apperception of the world changes, and correcting course as he perceives his errors."[51] Indeed, any interference from others that influences this "quality" of an individual must be considered immoral and highly dangerous.

Hence, the freedom to live an unwatched life relates to this problem of respecting people as choosers. I agree with Benn: spying is objectionable because it deliberately deceives a person about his world and, for reasons that *cannot* be his own, affects his attempts to make rational choices. It is impossible to respect someone as engaged in a worthy enterprise if we deliberately but secretly alter the circumstances in which he operates, concealing the fact from him. In this case, the offense stems from the spy's intrusion into the subject's life. The actions of the observer – person A – are likely to affect the project undertaken by the observed – person B – and these actions therefore change person B's perception of, say, an interaction with person C that occurred under A's watch. If B becomes aware of A's surveillance, B may regard a conversation with C very

[50] Benn, 1984, p. 227.
[51] Ibid., p. 229.

differently, perhaps no longer even recognizing it as the same enterprise. If B remains unaware of the intrusion, however, his perception of his exchange with C will be inaccurate, distorted. As Benn says, B has false notions about his life because he has assumed that his enterprise is unobserved: "He may be in a fool's paradise or a fool's hell; either way, A is making a fool of him. ... I can well imagine myself freely consenting to someone's watching me at work, but deeply resenting anyone's doing so without my knowledge – as though it didn't matter whether I liked it or not."

Person B is disrespected not only because his privacy is violated but also because he is inhibited from acting freely as a self-determining chooser: his actions should not be arbitrarily frustrated by a denial of privacy. Of course, privacy is also related to the ownership of our bodies and the right to maintain control over them.[52]

We can easily summarize this section: I am not obliged to agree with or support other human beings' projects, choices, and feelings, but I do have to respect them and recognize that other people should determine their own destinies through their own decisions as much as possible. I think that control over one's own destiny is one of the most important aspects of the modern conception of morality itself, and its lack – real or perceived – adds to the feelings of powerlessness and randomness in our lives.[53] It is only when our freedom is respected and our ability to take responsibility for our choices is guaranteed that we can exert more control over our lives and obtain what we deserve.

EQUALITY AND RESPONSIBILITY

While freedom and responsibility are usually thought of in the context of collectivities, the concepts also relate to our lives at a personal level. An interesting theory posed by Ronald Dworkin nicely demonstrates how the interplay between individual and institutional burdens relates to an imbalance in people's commitments and roles.

Following Dworkin's thinking,[54] responsibility is the basis for the equal distribution of wealth – the way financial resources are allocated depends, of course, on how certain laws control and direct ownership, theft, contracts, and torts as well as welfare, taxes, labor, civil rights, and environmental issues. In order governments to maintain equal concern for all of their citizens, Dworkin contends that leaders must seek a

[52] Reiman, 1984.
[53] On the negative aspects of privacy – nepotistic family solidarity, the concealing of crimes, and the abuse of women and children, for example – see Benn, 1984. On privacy and self-incrimination, see Gerstein, 1984.
[54] Dworkin, 1993.

material balance that he calls "equality of resources" rather than focus solely on the equality of welfare.

This focus on equal resources reveals the deep connection between political morality and ethics in general: we should not think of democracy and liberty as being in conflict with equality (as Rawls does, for example); instead, they fulfill ethical assumptions about the characteristics of a good life. Two principles of ethical individualism demonstrate the link between politics and morality, according to Dworkin. The first is the notion of equal importance: objectively speaking, it is important that every human life be successful rather than wasted. The second principle is the idea of special responsibility: "though we must all recognize the equal objective importance of the success of a human life, one person has a special and final responsibility for that success – the person whose life it is."[55]

The first principle holds that people have a moral obligation to act with "as much concern for the fate of everyone else in the world as for their own fate or that of their families and friends."[56] Political bodies must adhere to this idea; they are required to regard all citizens with impartial objectivity. The second principle deals with the idea of individual responsibility in that

> [s]o far as choices are to be made about the kind of life a person lives, within the range of choices permitted by resource and culture, he is responsible for making those choices himself. The principle does not endorse any choice of ethical value. It does not condemn a life that is traditional and unexciting, or one that is novel and eccentric, so long as that life has not been forced upon someone by the judgment of others that it is the right life for him to lead.[57]

The two principles are closely related: they have to act in concert so that the first compels governments and other institutions to adopt laws and policies that ensure that their citizens can, as I say, control their destinies responsibly. This can happen only when governments and collective institutions are able to make people "insensitive to who they otherwise are – their economic background, gender, race, or particular sets of skills and handicaps."[58] Governments and collective institutions, if they can achieve this result, can make people's destinies sensitive to the choices they make and have made, which, in turn, allows these same people to act in more productive ways, as required by the second principle.

[55] Dworkin, 2000, p. 5.
[56] Ibid.
[57] Ibid., p. 6.
[58] Ibid.

This ethical and political construct ensures that we take responsibility for consequences arising from our own choices – choices we make, of course, based on our own beliefs and preferences. Dworkin adds, however, "I make no assumption that people choose their convictions or preferences, or their personality more generally, any more than they choose their race or physical or mental abilities,"[59] and he goes on to point out some of the limitations of conservative and egalitarian politics:

> The old egalitarians insisted that a political community has a collective responsibility to show equal concern for all its citizens, but they defined that equal concern in a way that ignored the citizens' personal responsibilities. Conservatives – new and old – insisted on those personal responsibilities, but they have defined them so as to ignore the collective responsibility.[60]

Of course, we should add that it is impossible to take responsibility for our choices if they are not freely made – that is, if they are distorted by bad faith, as illustrated earlier, or if they are dictated or manipulated by others. Successfully owning one's own destiny, then, depends not only on the individual in question, but also on the collectivities surrounding him or her. In order to ensure that people assume responsibility for themselves, governments and communities must implement policies that support free choice and individual agency and reduce our vulnerability to luck and randomness.

So an individual's commitment is critical, but the collective's responsibility to all people it affects is equally important: "When injustice is substantial and pervasive in a political community, any private citizen who accepts a personal responsibility to do whatever he possibly can to repair it will end by denying himself the personal projects and attachments, as well as the pleasures and frivolities, that are essential to a decent and rewarding life."[61] When we are able to acknowledge the responsibility of both the individual and the collective, the interconnectedness of democracy, liberty, and equality becomes very clear.

IMMORALITY AND LACK OF MORALITY

Certainly a lack of knowledge, whether through ignorance or repression, is responsible for the condition of bad faith and, as I have described in the previous chapters, for many other unethical outcomes. I suspect a lack of knowledge is also a root cause of immorality. This brief section aims to

[59] Ibid., p. 7.
[60] Ibid.
[61] Ibid., p. 236.

briefly address some problems related to immorality, to its creative character, and to the concept of "immoralism."

Certainly immorality is thought of as an absence of morality; the potential for immorality grows whenever there is any decrease in freedom and responsibility. When faced with immoral outcomes, people often say, as Mill puts it, "it was impossible to choose, it was impossible to intervene." But, he continues, "a person may cause evil to others not only by his actions but by his inaction, and in neither case is he justly accountable to them for the injury."[62]

As I will illustrate in the next chapter, I think morality is closely connected to something "active" and, in some cases,"creative." But one can also hypothesize a similar link when it comes to immoral action, which is not merely the failure to follow moral rules – it must be constructed. I think that immorality is not necessarily the opposite of morality; it arises from very specific kinds of reasoning, knowledge, and feeling that are not simply the negative side of already established moral traditions.

For some intellectuals, the idea of morality is distasteful. I too think that there is indeed something slavish and weak, as Nietzsche maintains, in the drive to "teach" people what to do, which, of course, inhibits both desire and the radical idea of freedom.[63] If we put a post-modern twist on Kant (who, as we've seen, saw ethical knowledge as something simple and readily available to all people of good will), it also seems that, as Bruno Latour says, "If there is one thing that does not require an expert, and cannot be taken out of the hands of the ten thousand fools, it is deciding what is right and wrong. ... Yes, morality is a phantom of statesmanship."[64]

If only we could avoid the problem of morality entirely, or, at the very least, adopt simplistic, childish attitudes toward it that are less taxing to maintain! For those who yearn for guidance, Mill's writings are both helpful and unassailable. In *On Liberty*, he writes the following:

> The object of this Essay is to assert one very simple principle, as entitles to govern absolutely the dealings of society with the individual in the way of compulsion and control, whether the means used be physical force in the form of legal penalties, or the moral coercion of public opinion. That principle is, that the sole end for which mankind is warranted, individually or collectively in interfering with the liberty of action of any of their number, is self-protection. That the only purpose

[62] Mill, 1966, p. 17.

[63] While immoralists like Nietzsche hold that moral reasons are not necessarily good reasons, they need not claim that there are no good reasons for supporting many of the norms that are accepted as moral norms. Immoralism simply seems to declare that life would be better without guilt and blame and to deny that these are the only means for encouraging the discipline necessary for an ethical life.

[64] Latour, 1999, p. 243.

for which power can be rightfully exercised over any member of civilized community, against his will, is to prevent harm to others. His own good, either physical or moral, is not a sufficient warrant.[65]

This basic tenet relates not only to morality but also to the prosaic yet essential fact that "[a]ll that makes the existence valuable to any one, depends on the enforcement of restraints upon actions of other people."[66] Freedom and its exercise are important – which is why we wish to avoid bad faith, of course – but, as we will see, restraints are necessary, too.

GLOBALIZATION, COMMODIFICATION, AND WAR

As we have seen, "being-for-others" is, for one in bad faith, an external instrumental condition that does not exhaustively define one's own identity. In our technological world, "being-for-others" is a particularly intense position to be in because of our status as hybrid people, which I illustrated in chapter 3. The blurring of traditional distinctions between people and things that makes us a kind of cyborg is certainly created not only by our relationships with technological products, but also by the externalization we achieve through our intertwining with other external artifacts, like institutions, roles, and social duties. These aspects are connected to the increasing "commodification" of our lives, a trend that is multiplying the potential situations in which we function as cyborgs, and it is therefore useful to briefly analyze how commodification, wrought mainly by modern globalization, is affecting our lives.

We know that many modern languages are being reduced to "kitchen languages"[67]; that is, they are increasingly relegated to the home and social situations and generally eschewed in work settings. This is becoming the fate, for example, of my first language, Italian, which Galileo used for political and epistemological purposes in his early modern writings: he defied tradition by using that language to invent modern science at a time when Latin was the expected language of scholars. The use of Italian and many other languages is declining, as everyone knows, because the inexorable process of globalization is establishing one dominant language – English. This shift, however, is just one effect of globalization, whose repercussions reach far beyond the way we communicate with each other.

The condition of human beings in the globalized world is depicted in similar ways by many authors,[68] and while the descriptions vary in

65 Mill, 1966, pp. 14–15.
66 Ibid., p. 10.
67 Teeple, 2000, p. 16.
68 Sachs, 2000.

political slant, they are surprisingly consistent in actual content. In these writings, the new world is afflicted by powers of oppression and destruction, but it also features new possibilities and opportunities for humanity. It seems that neoliberal policies seek to create a global system of internationalized capital and supranational banking networks and institutions controlled by hegemonic corporations and maintained by the free movement of capital across national borders: as Teeple writes, "The formation of cartels, oligopolies, or monopolies to control supply and demand, geographic markets, and prices; and the growth of the advertising sector, which is in effect an attempt to control and create demand," reflect corporate planning that "is simply the recognition that the unreliability of the market ... cannot be tolerated given the enormity of capital investment."[69]

In the era of globalization, all people are welcome as potential contributors regardless of race, creed, color, gender, sexual orientation, and so forth. Nevertheless, it seems that, at least since the late 1970s, this new international system has brought with it greater economic inequality and that, as a result, many global problems are worsening: low wages, unemployment, illiteracy, poverty, child labor, forced emigration and transmigration, forced labor, the social and professional subordination of women, war, slaughter, bribery and corruption, and disease and morbidity.

The list of globalization's negative effects goes on and on: pollution and environmental destruction seem out of control, as we've mentioned before; the sovereignty of the old nation-states is declining; labor and trade union rights, health care, social assistance, support for the elderly, and educational standards are all declining; legislative assaults on wages are being launched by governments around the world at a time when Keynesian welfare reform has become less popular; corporations control most mass media outlets, which negatively affects local cultures;[70] the increasing privatization of public services like education, health care, and pension plans is bringing few benefits for the young, the ill, and the elderly; state funds are more and more often redirected to the private sector (take the American school voucher system, for example); human and civil rights[71] as well as liberal democracy are circumscribed and threatened; and the old national strategies to assist developing nations

[69] Teeple, 2000, p. 37. Rajan and Zingales, in their recent *Saving Capitalism from the Capitalists*, describe how strong economic groups tend to hinder capitalistic competion, taking advantage of their leading position in the market.

[70] According to the French antropologist Jean Loup Amselle (2001), the process of globalization would not destroy all local cultures. On the contrary, it could uncover the existence of cultures that otherwise would remain unknown.

[71] McGinn, 2000; Dyson, 2000.

seem to have been destroyed. Finally, a growing sense of disillusionment and cynicism is affecting people of all political tendencies.[72]

There is a general crisis regarding established institutions – for example, many believe that, as its worth is questioned and its effectiveness challenged, the nuclear family is under threat: "Children, moreover, remain by and large the property of their parents or wards of the state, poorly protected by civil rights and largely unrecognized and unseen as embodiments of humanity, but they are increasingly made the objects of consumer marketing."[73]

The global market is the new reality, together with the globalization of production, distribution, and exchange. Unlike the United Nations, the new transnational institutions are primarily economic or commercial in nature; as such, they lack democratic and political legitimacy because they are not products of free elections,[74] yet they wield enormous power in the world: the result is, as Rosenau has put it, "governance without government."[75] Those who work to counter the effects of such institutions are few in number and limited in their aims (and unfortunately their behavior is not always transparent) – I'm thinking here of such collectivities as religious and ecological organizations, Aboriginal alliances, consumer protection groups, old age advocacy coalitions, civil liberty associations like Amnesty International, women's movements, antinuclear groups, and organizations like Oxfam and Médecins sans Frontières.[76]

Globalized human beings seem disenfranchised: because they are fragmented – paradoxically, their communications are obstructed in this era of heightened communications – they cannot be represented in the global theater.[77] Unscrupulousness at both village and global levels exacerbates their basic segmentation and renders them more and more marginalized. Corruption, disease, and frustration begin to take the form of psychosis, substance abuse, anguish, and boredom.[78]

Moreover, human beings' activity and labor have shifted from energy-intensive to information-intensive; information is increasingly objectified in computers and networks and so is more often located outside of human carriers. Economic control is not merely control of a sector of human life that can be separated from the rest; it now seems to permeate all aspects of all of our activities.

72 Teeple, 2000, pp. 2–4.
73 Ibid., p. 41.
74 Schiller, 2000.
75 Rosenau, 1992.
76 Pech and Padis, 2004, shed light on some controversial aspects of the financial management of the so-called NGOs (nongovernmental organizations).
77 Teeple, 2000, p. 197.
78 Hardt and Negri, 2001, pp. 389–392.

For example, let us draw from the topic of biotechnology in chapter 2 and consider the case of computational and informational "nonhuman" objects and tools. An increasing number of institutions have been transformed into "virtual" entities: banks, for example, corporate headquarters, government offices, universities and schools, health care organizations, and the entertainment and advertising industries. Businesses have made it possible to buy airline tickets and other goods from online sources. In all these cases, computer systems store, distribute, transform, and apply information.

Human beings have been excised from many transactions – economic and otherwise – as the tasks they once managed have been transferred to external things like computer systems and networks. It seems that many professionals have been affected by this process: certainly in fields such as medicine, law, engineering, architecture, and teaching, human beings are embodiments of specialized accumulated knowledge, and, as a result, they serve as "biological" repositories, disseminators, and processors. The current trend, however, is to fill these roles, many of which require significant skill, with nonhuman computers and other tools. This movement signals a kind of "demise of the expert," with the term "expert" conveying the idea of knowledge monopolies held by members of particular groups. It is true that technology has loosened the grip on certain information once held by various professions and nations, but, at the same time, an increase in the number of patents and intellectual copyrights means that corporate monopolies are growing.

While globalization's negative effects are widely known – the subordination of local cultural traditions to large-scale market and corporate interests, for instance – I contend that this new era of locating knowledge outside human carriers also brings with it the potential for at least some good. As knowledge and skill are objectified in nonhuman mediators (things that start to think and things that make us smart),[79] outside of human carriers, many positive outcomes become possible: (1) the democratizing and universal dissemination of knowledge; (2) greater ownership and wider transmission of information once controlled by corporate monopolies; and (3) less emphasis on labor as the source of value, which would transform the relationship between labor and capital.[80] Globalization's tendency to shift knowledge to nonhuman repositories could be beneficial, for, in so doing, it makes information universally accessible. A greater pool of available knowledge could lead to interesting new possibilities while enhancing freedom and increasing free choice.

Finally, some authors maintain that the era of globalization presents an increasing and all-encompassing commodification of sociocultural needs,

[79] Cf. Gershenfeld, 1999, and Norman, 1993.
[80] Teeple, 2000, pp. 70–71.

that is, of human "cultures," features, and actions.[81] Many subjectivities have become more and more enmeshed with economic relationships; almost all aspects of our lives and the entire realm of reproduction, for instance, are influenced by economic transactions. In this light, the ethical problems of market rhetoric – "partial alienability" and "market inalienability" (aspiring to noncommodification) – become particularly important when it comes to human endowments. Are babies, human organs, blood, labor, fetal gestation (surrogate motherhood) alienable or not? What about sexual services, genetic enhancement, cloning, and bodily integrity? Finally, can we alienate the traditional liberal life-liberty-property triad or our voting rights?[82] To think that something personal – a right or attribute, say – is fungible implies that it is "separate," and thus its owner is dichotomized and alienated.

I agree with Radin's comment: workers who adopt market rhetoric dichotomize their own labor as a commodity and themselves as persons; they dissociate their daily life from their self-conception. On the other hand, workers who do not consider their labor a commodity are alienated from others who do, because, from the workers' perspective, people who conceive of their labor as a commodity fail to consider themselves as whole persons: "To conceive of something personal not fungible also assumes that persons cannot freely give of themselves to others. At best they can bestow commodities. At worst – in universal commodification – the gift is conceived of as a bargain. ... Commodified sex leaves the parties as separated individuals and perhaps reinforce their separateness. ... Noncommodified sex ideally diminishes separateness; it is conceived of as a union because it is ideally a sharing of selves."[83]

In some cases, the increasing commodification of human aspects and actions generates uncertainty. In a divorce, for instance, how is the monetary value of a spouse's professional degree or work as a stay-at-home parent decided? Does rape damage bodily integrity in ways that can be economically quantified? When human beings' features are alienable, we immediately think that they have become a "means." Imagine a person who has decided to sell sex so that he or she can be a "means" for other people. A society may consider this behavior immoral, but this does not imply that the sexually commodified body of that person should be less respected or that his or her problems are less worthy of moral consideration. The concept of "respecting people as things" also provides an ethical framework that allows us to interrogate and analyze the condition of modern people and then develop ways to

[81] Postman, 2000.
[82] Cf. Posner, 1981, and Radin, 1999.
[83] Cf. Radin, 1999, p. 122.

support them – even as nearly all parts of their lives are increasingly commodified.

COMMODIFICATION OF HUMAN DIGNITY?

In this time of increasing technology, a time when we seem to be commodifying nearly everything in sight, is it possible to turn even intangible values like human dignity into commodities? The prospect sounds ominous, but I believe doing so could actually bring about positive change; attaching economic worth to human dignity could generate a certain degree of social demand and need for it. Indeed, in our current collectivities there is already a call for improvements in human dignity, and its commodification could serve as a beachhead for those working to spread dignity. Take, for example, a university that offers special support to students (more tutoring, more capacity to listen to the students' problems, more democracy and participation, etc.) or to its teachers (more time for them to do research, more latitude in deciding whether to teach a course,[84] etc.); such an institution could use economic transactions to enhance the lives of its students and faculty members. There could be an option for families and parents to pay more for a level of treatment that ensures greater dignity for both students and teachers, which would create an environment that would attract higher-quality instructors and generally enhance the reputation of the institution. Similar arrangements could be implemented in hospitals and other enterprises. But we would need to consider and manage the downside of such a strategy – namely, that such a system would allow the wealthy to buy even greater dignity than they are already accorded!

Of course, many new links between commerce and ethics have already been established – companies concerned about sustainability are marketing ecological products,[85] and so-called ethical banks, which finance humanitarian endeavors, are inviting investments that in turn help to fund organizations that care for the elderly, treat drug addicts, work for ecological improvement, and so on. "Dignity" products could be clearly

[84] On one occasion a university "respected" me as a teacher and as a person in the following way: its bureaucracy asked me by e-mail at 1:26 P.M. to decide whether I was able to accept a visiting professorship; the deadline was 5:00 P.M. on the same day. I read the message at about 3:30 P.M., so I had to decide in an hour and a half! I contend that giving people adequate time to make decisions is a form of respecting them. Only in this way are people really able to choose freely and, therefore, to take full responsibility for their decisions (for example, by gathering and checking the information needed to make the decision).

[85] For example, those not related to environmental waste or to the exploitation of child laborers, etc. See also chapter 1 of this book, the section "Preserving Things: Technosphere/Biosphere, Human/Nonhuman."

labeled as such and promoted, marketed, and sold as items that enhance human lives. Along those lines, some products can now become "fair trade certified," a designation given by one of the nineteen members of Fairtrade Labelling Organizations International. These groups audit transactions between corporations in wealthy countries and suppliers in developing nations; for a company to earn and retain certification, it must demonstrate that it has paid farmers and other workers a fair, above-market price. In some American supermarkets, for example, it is now possible to find fair-trade-certified bananas and coffee as well as products like Better World Hot Cocoa™. These items are typically more expensive than their conventional counterparts, but we must remember that inhabitants of wealthy countries often enjoy bargains because workers in poor countries are underpaid. Spending a little more money for a bunch of certified bananas means that the farmers who grew them have been paid wages that allow them to support themselves and their families.

Some aspects of human dignity are so clear and easy to appreciate that it seems quite feasible to create market fungibility for products that promote it. Moreover, if we work harder to commodify aspects of human dignity, we can also counterbalance the current negative effects of commodifying other human aspects, many of which I have illustrated already in this chapter.

"RESPECTING PEOPLE AS THINGS" IN WAR

When treating the problem of ecotage and monkey-wrenching in chapter 1 ("Ends Justify Means"), I emphasized the urgency of defining some "things" as distributed "moral mediators"[86] that can provide precious ethical information not attainable through existing internal mental resources and ascribe important new values to human beings that could not otherwise be ascribed. I observed that in radical environmentalism, ecotage is seen as a way to defend minorities, and that unprotected plants and animals that are being destroyed are considered, in this construct, minorities. Moreover, if the self is ecologically extended, as many scholars contend, and is part of a larger ecological self that comprises the whole biological community, then the defense of nonhuman "things" is simply "self-defense." Consequently, breaking the law is justifiable for some activists.

A similar line of thinking can help us to reframe our understanding of modern warfare and the threat it poses to nonmilitary entities. In some contemporary wars waged by Western countries, military forces have systematically destroyed ill-fated sets of local things, animals, and

[86] Cf. chapter 6.

human beings, all of which have no defenses of their own and are rarely regarded by those in power as having enough worth to warrant protection. Paradoxically, they appear on the world's moral radar screen only because of the ethical value they acquire *after* being destroyed: unfortunately, dead bodies and bombed-out buildings have a particular kind of worth that is unavailable to living people and unscathed cities.

I contend that the traditional ethics of war has not paid sufficient conceptual and strategic attention to the problem of noncombatant immunity. This lapse has occurred, I believe, because a distorted way of "respecting people as things" has led us to assign exaggerated ethical and pedagogical value to dead bodies and destroyed objects. Ironically, in war human beings are morally "respected" as people only when they become nonliving things – that is, when they are dead – just as certain animals and sites gain value when they go extinct or are destroyed. In a way, so-called smart bombs seek to minimize this bizarre "respect for people as things" by targeting only military sites and limiting civilian casualties.

I am not concerned here with the concept of a just war or the question of "when to fight"[87]; rather, I would like to focus on the problem of "how to fight" by revisiting the concept of "respecting people as things." The generally accepted principle of noncombatant immunity is that civilian life and property should not be subjected to military force, but there are conflicting ideas about what counts as "civilian" and what counts as "military." I believe that we can consider unprotected things, animals, and human beings that are threatened by warfare to be minorities, just as environmentalists consider endangered plants and animals to be minorities; this status would give them noncombatant immunity and accord them protection. The problem of "how to fight" would, consequently, acquire new variables and moral nuances and so help us to recalibrate the ethics of war.

Those who refuse to acknowledge the potential for collateral damage create for themselves a sort of psychological refuge from the horrors of war; they find relief in such denial, but it also leads them to underestimate the problem of noncombatant immunity. This brings us back to the issue of bad faith: building an emotional firewall is a way for an individual to construct another self, one less sensitive to the horrors of war. War kills human beings, and this fact is too horrific for some people to confront squarely; such people – politicians, sometimes – cosset themselves in bad faith rather than sorting through all the complexities of war. By so doing, they maintain their ignorance about the problem of noncombatant immunity, and it is this ignorance that perpetuates

[87] A complete treatment of the ethics of war and peace is given in Lackey, 1989.

human anguish. In turn, more human suffering drives more people to the opiate comfort of the condition of bad faith, and the cycle continues unchecked.

Moreover, convincing oneself that collateral damage is unlikely (or, worse, acceptable) contributes to an objective ideology about war that is available "out there," stored in external devices and supports (other people, books, media, etc.) of a given social collective. People readily pick up external ideological "tools" of this kind, then re-represent them internally in order to preserve their bad faith. The condition of bad faith may become so widespread in a cultural collective that it becomes crystallized in ideological narratives shared by entire communities: bad faith becomes, in essence, a culture's default setting. Only by continually monitoring the links between the internal and external can we lay bare the deceptive nature of our beliefs. This process testifies to the continuous interplay between internal and external, and I think it can nicely illustrate the deceptive character of the various ideologies.[88]

Wars compel a culture to acknowledge the fact that it attributes greater value and respect to tanks and technological weapons and the all-encompassing commodification of human needs than it does to an intact natural community of living plants, animals, and human beings. Considering how entrenched many old ideas are, making the case for new kinds of war or nonmilitary strategies to achieve prosperity and freedom is a considerable challenge indeed.

Respecting people as things is important not only for ascribing greater value to others but also for preserving our own dignity and freedom. As we have said, embracing this rather un-Kantian concept requires that we develop new ways of thinking, new kinds of knowledge; if we fail to do so, one of the negative consequences is that we put ourselves at greater risk of stumbling into bad faith.

To live without important knowledge – whether we have resisted it or are unaware of it – is to create for ourselves a toxic state of ignorance that generates a vague sense of anxiety, which, I contend, is what drives us into the bad faith condition. We wriggle away from unpleasantness and retreat into a kind of oblivion that only *seems* less harmful than confronting difficult issues: bad faith ultimately has a deeply corrosive effect on our well-being, for being ignorant of possible choices constricts our freedom and diminishes our dignity.

But this invisible condition is so pervasive, so much a part of life in the modern world, that it is not easy to imagine how to rise above it. Bad faith may be a self-defeating coping mechanism, but it is, for many people, the

[88] On the relationship between these ideologies about wars and the paroxysm of violence in sacrificial-religious-victimary processes, cf. Girard's (1979) seminal work.

best (or only) one available. The answer is to supplant it with knowledge; that which seems unbearable or impossible can often, when one has the right information, be approached differently and managed, if not easily, then at least more comfortably. Here we return to the challenge of building new forms of knowledge that, once constructed, we must use wisely and purposefully. Acquiring a new understanding of science, technology, and ourselves is critical, but it does not complete our quest to enhance human dignity. Once we have this new awareness, we face the practical matter of putting it to use. Therefore, the next step in the process is to identify the principles that will guide us and to develop reasoning strategies that will lead us to the best choices possible.

6

Creating Ethics

Good Reasons and Good Arguments

> Morals excite passions, and produce or prevent actions. Reason of itself is utterly impotent in this particular. The rules of morality, therefore, are not conclusions of our reason.
>
> David Hume, *A Treatise on Human Nature*

While I deeply believe that creating and acquiring new knowledge is critically important, even I must admit that all the information in the world is meaningless unless we can use it effectively: the principles and ways of reasoning that allow us to put new ethical knowledge to work are just as important as the knowledge itself. The remainder of this book, then, will have a methodological focus: while the next chapter employs an epistemological and cognitive approach to explore the basic aspects of abductive scientific reasoning, the present one will consider its moral side. Hence, we can say that these last two chapters of the book are in some way twins, because they systematically treat similar issues in complementary ways. In using this approach, my hope is to reorient and modernize philosophical discussion in a way that eliminates the burden that traditional philosophy puts on moral agents by requiring an unrealistic level of competence, information, and (formal) reasoning ability.

Thus far, we have considered how new technology and changing economic circumstances have generated the need for new moral knowledge, and we have examined the social and cultural contexts in which moral reasoning takes place. Yet to be accomplished, however, is an analysis of moral reasoning's cognitive features and constraints, and to that end, this chapter will explore how recent research in epistemology, cognitive science, artificial intelligence, and computational philosophy offers new ways for us to understand ethical knowledge and reasoning.

I believe that we can find models for practical reasoning in scientific thinking and problem solving – an appropriate source given the fact that science and technology underlie many of the cultural changes that have triggered the need for a new approach to ethics. It is worth remembering here that ethical knowledge, like scientific and other kinds of knowledge, is created and used by human beings and is, therefore, fundamentally related to several cognitive and epistemological concerns. First of all, I will show that ethical knowledge and reasoning are expressed not only in words at a verbal/propositional level but also through model-based and manipulative/"through doing" processes. Consider, for example, the important role in ethics played by imagination, which, like analogy, visualization, simulation, and thought experiments, is a form of model-based reasoning.

Another important theme is creativity, which is an important factor in effecting conceptual change and forging new perspectives. By exploiting the concepts of "thinking through doing" and of manipulative abduction, I will also illustrate some of the most interesting cognitive aspects of what I call "moral mediators." Morality is often enacted in a tacit way – so to speak, "through doing" – and part of this "doing" can be seen as a manipulation of the external world with the express purpose of building these moral mediators. They can be built purposefully to achieve ethical effects, but they may also exist in a variety of entities independently of human beings' intentionality and may carry ethical or unethical consequences. Furthermore, while describing morality "through doing" I have supplied a list of "moral templates" – forms of invariant behavior – that illustrate manipulative ethical reasoning.

My analysis of moral knowledge and inference in the cognitive terms of abduction and model-based reasoning yields a very useful integrated framework that reveals connections between aspects of ethical deliberation that are typically considered dissimilar if not wholly contrasting – from the role of emotions, visualizations, and narratives to the function of schemas in moral thinking and deliberation. Finally, the concept of abduction has also allowed a cognitive comparison to diagnostic reasoning that brings to light the intrinsic "incompleteness" of knowledge in ethical inferences.

GOOD REASONS AND GOOD ARGUMENTS

The minimum conception of morality is very concisely described by James Rachels: "Morality is, at the very last, the effort to guide one's conduct by reason – that is, to do what there are the best reasons for doing – while giving equal weight to the interests of each individual who will be affected by one's conduct: there are no privileged people."[1]

I would add that to achieve even this minimal level of morality, we need guidelines and reasoning skills that allow us to employ our

[1] Rachels, 1999, p. 19.

knowledge effectively: (1) we need sound principles for choosing actions, principles that allow us to consider opposing views, and (2) we need appropriate ways of reasoning – inferences – that permit us to apply these principles. The first statement relates to the concept of "knowledge as duty" from chapter 4: it is our duty to establish principles – that is, to produce and apply rich ethical knowledge appropriate to addressing modern moral problems, just as it is our responsibility to seek and use other kinds of knowledge – "scientific" knowledge, for example.

The second statement, on the other hand, stresses the importance of being methodologically aware of moral reasoning's cognitive processes. Dealing dexterously with modern moral challenges requires new knowledge, of course, as well as greater skill in acquiring and understanding pertinent facts and information. As I have illustrated in chapter 3, I maintain that both appropriate ethical knowledge and proper moral reasoning are necessary if human beings are to assume responsibility and enjoy freedom, both of which, as I argued in the previous chapter, are primary features of a moral life.

How we ought to live is the central problem of morality, and the way we reason dictates how we live. Our intellectual and religious traditions have elaborated many ethical theories, each of which seeks to answer this question. Ethical theories are formed by various moral rules and principles that aim at providing good reasons to guide us when making inferences and moral judgments. Many kinds of reasoning are employed in ethics, both in the construction of theories and knowledge and in practical deliberation. Consequently, *creating ethics* refers not only to expanding *moral knowledge* but also to developing and using *moral inferences* systematically when dealing with problems, real or abstract, in practical deliberation.[2]

Generally speaking, living morally involves matching a concrete situation or problem with something abstract – rules, principles, and prototypes, for example, as well as models, emotions, and so on. Principles in ethics are easy to identify; prototypes (like schemas and frames) are standard, previously solved ethical cases or common situations involving ethical issues.[3] In the following sections, I will consider other strategies for ethical deliberation, like the uses of models, emotions, and what I call "moral mediators," which, until now, have been neglected in moral philosophy.

[2] An amazing architecture for building an ethical robot (both deontological and consequentialist) is illustrated in Gips, 1995.

[3] Cf. the section "Model-Based Reasoning" later in this chapter. Ethical prototypes have been analyzed from a cognitive perspective by May, Friedman, and Clark (1996). An interesting approach to moral knowledge and reasoning in terms of prototypes and parallel distributed processing is presented in Guarini, 1996.

Living morally is not always simple, especially in the face of daunting new challenges that require creative new ideas: for example, as we discussed in the first four chapters, certain technological innovations have generated problems in the realms of human reproduction, the environment, and cyberspace that were unimaginable just twenty years ago. If we are to succeed in managing these and other unprecedented difficulties, we must use novel ways of thinking and feeling to recast and reinterpret the situations that have created them. Without adequate reasoning, even well-intentioned moral actions may fail – or, worse still, cause harm – and the best way to facilitate adequate reasoning is to confront problems with flexible and well-fed minds. As we will see later in this chapter, the level of ethical creativity, for both collectivities and individuals, is directly related to the richness of moral knowledge and the quality of emotional training combined with the strength of will to do the right things.

From a theoretical point of view, living morally is the capacity to use valuable moral knowledge along with well-developed *templates* of reasoning that can explain situations or conflicts and provide suitable modes of deliberation. Long-term pondering can thwart the process: sometimes immediate action is critical – one must quickly manipulate objects and situations in the human and natural environment (cf. the section later in this chapter "Being Moral through Doing: Taking Care"), and in such instances morality is achieved "through doing" rather than "through feeling" or "through reasoning."

CREATING AND SELECTING REASONS

A famous case can clearly show how different theories (principles, rules, and prototypes) and inferences lead to different conclusions in ethical deliberations. The case, reported by Rachels,[4] involves an infant called Baby Jane Doe who was born in New York in 1983. The baby suffered from multiple birth defects, including spina bifida (a condition in which an embryo's spinal column fails to close completely during the first month of development), hydrocephaly (excess fluid on the brain), and microcefaly. Baby Jane Doe needed surgery to close the gap in her spinal column, but her pediatric neurologists disagreed on the potential outcome of the procedure: one physician, Dr. Newman, believed the surgery would not be worth undertaking because the baby's defects were so great that even with the operation, she could never attain a reasonable quality of life, while the other physician, Dr. Keuskamp, advised immediate surgery because he thought the baby's condition was less dire. The parents, faced with this medical *and* moral problem, followed Dr. Newman's advice and refused surgical treatment. Shortly after that, Lawrence

[4] Rachels, 1999, pp. 6–12.

Washburn, a conservative "right-to-life" lawyer, petitioned the courts to set aside the parents' decisions and order the surgery. The New York State Supreme Court granted that request, but a higher court, calling Washburn's suit "offensive," quickly overturned the order after hearing the description of the baby in Dr. Newman's testimony:

> ... on the basis of the combination of malformations that are present in this child, she is not likely to ever achieve any meaningful interaction with her environment, nor ever achieve any interpersonal relationships, the very qualities which we consider human.[5]

At this point the federal government got into the act by asking whether a "handicapped person" – the infant – was being discriminated against. Still, based on Dr. Newman's prediction, the suit to order surgery was dismissed, and consequently the procedure was not performed.

It is clear that Newman's testimony was based on a kind of *predictive reasoning*. This kind of reasoning, if combined with the ethical principle of *benefit*, states that if we can benefit someone without harming anyone else, we ought to do so, and it was this thinking that led to the conclusion that surgery would not be worth the risk and expense: "With surgery, she would have a 50–50 chance of surviving into her 20s, but she would be severely mentally retarded, paralyzed, epileptic, unable to leave her bed, without control of her bladder or bowels, and unusually vulnerable to such further diseases as meningitis. The mental retardation would be so severe that she would never even be able to recognize her parents."[6] Surgery, then, would benefit neither Baby Jane Doe nor anyone else. Even if she survived the operation, her parents would face years of heavy labor caring for a child who had derived little benefit from it.

If, however, instead of the degree-of-benefit principle, we prefer to use the *sanctity-of-human-life* guideline followed by right-to-life activists, we would view the surgery very differently. In the construct of this second principle, which states that every human life is invaluable and sacred, the pessimistic prospects determined through predictive reasoning are irrelevant, and we must choose surgery, regardless of the possible outcome. Of course, we would arrive at a similar conclusion if we applied the principle of the *wrongness of discriminating against the handicapped*. The failure to perform surgery on the baby would, in this case, be considered unacceptable discrimination against, and a denial of the rights of, a handicapped person.

When considering the problem of Baby Jane Doe, we have at our disposal three ethical principles to guide us – degree of benefit, sanctity of

[5] Ibid., p. 7.
[6] Ibid., pp. 7–8.

human life, and wrongness of discriminating against the handicapped – which can be chosen from a preestablished list of principles ("reasons") that guide us as we evaluate problems and arrive at decisions. I contend that ethical deliberation as a form of practical reasoning[7] is a variety of *abductive reasoning*, similar to the reasoning conducted in scientific and diagnostic settings.[8] Of course, in a moral case we have reasons that support conclusions instead of explanations that account for data;[9] still, moral reasons can play an important explanatory role, as we will see.[10]

We have said that moral deliberation involves selecting or creating "reasons" and then applying them to concrete cases. We *select* (or *create*, if we do not have any) moral reasons and then apply them to concrete cases, sometimes while also looking for the "best" reason according to some ethical meta-criterion. When we create new ethics, we produce new knowledge and form new rules about problems and situations that have not yet been fully interrogated from the moral point of view: in short, we construct new "reasons." We must not only use currently available ethical concerns/reasons to solve ethical problems but also build a richer body of moral knowledge in order to handle puzzling situations. In addition, new reasons (for example, in terms of new principles) tailored to the modern era will allow us to probe the moral intelligibility of problems in a fresh, original way. An example of such invention might be the relatively new ethical principle of civil disobedience, which provides "reasons" for some particularly extreme behaviors and therefore renders them morally intelligible. In the previous chapters, we have illustrated many new situations that require fresh moral knowledge – from the environment to cloning, from cyberprivacy to the problem of freedom and responsibility in the era of technology and globalization.

Nevertheless, in ethical deliberation we typically do not create new moral guidelines; rather, we select them from our "encyclopedia" of preexisting principles, as we did when we explored the Baby Jane Doe case from three established ethical viewpoints. From the sanctity-of-human-life perspective, all surgical operations that protect human life of any kind are good; surgery would also be the "right" choice if we assume that it is wrong to discriminate against a handicapped person by denying him or her treatment that someone without disabilities would ordinarily receive. On the other hand, if we, like the baby's parents, select the utilitarian principle of *benefit* from our encyclopedia ("If we can benefit

[7] Cf. Millgram, 2001 and 2001a.

[8] Magnani, 2001a.

[9] Cf. the following chapter, the sections "The Logical Structure of Reasons" and "Abductive Reasoning and Epistemic Mediators."

[10] On the various roles played by abduction in practical reasoning, cf. Gabbay and Woods, 2005, p. 50.

someone, without harming anyone else, we ought to do so"), we would opt to forgo the operation: based on the information available about Baby Jane Doe – that her chances of survival after surgery were poor and, even if she did survive, the prospects for her general health were grim – the surgery would offer little benefit to the infant and would not serve anyone else's interests. (See the later section "Expected Consequences and Incomplete Information.")

Even without the surgery, however, Baby Jane Doe did not die, and five years later she was doing much better than expected: she was able to talk, use a wheelchair, and attend a special school. Even though her parents had used a good utilitarian moral principle and the best information at hand when making their decision, their choice turned out to be "wrong" in spite of the fact that it was the product of careful moral consideration. This does not mean that the decision was immoral – it was just wrong. Other parents in other cases would be right to make the same decision.

Moral Progress

When confronting a moral issue, we have seen that we need not only ethical knowledge but also skillful reasoning if we are to render this knowledge applicable. In addition, we also stressed the fact that we often need new moral knowledge when faced with difficult new situations and tasks. These considerations can help in outlining moral progress. The notion of a universal truth in ethics is certainly a myth; so too is the idea of universal criteria that can always discern the "best" ethical theory or choice. But even though there is no one-size-fits-all moral principle, we should still strive to deal with unprecedented situations and problems by creating new ethical knowledge – that is, by accepting that knowledge is a duty.

New knowledge, for example, is one way to dispel the belief that some moral attitudes are better than others. For instance, take the developed Western world's objection to the widespread practice of human female castration; we assume that our condemnation is not just emotional, that it is instead a product of our sophisticated knowledge and reason rather than of our mere cultural traditions about sexual behaviors, social and economical relationships, families, and so on. In this way we can justify – in extreme cases – our avoiding the obligation to understand and respect others' cultural and social practices and relieve ourselves of the burden of cultural relativism.

EXPECTED CONSEQUENCES AND INCOMPLETE INFORMATION

"Knowledge as duty" naturally intersects with the central problem of "incomplete" knowledge. Spotty knowledge, however, is unavoidable: all ethical deliberations are based on intrinsically incomplete information because it is impossible for anyone to be aware of every fact related to any given subject. How, then, should we confront moral problems given that we are inescapably handicapped in virtually every situation by a lack of knowledge? I argue that we should confront difficult issues by exploring ethical reasoning through its verbal-propositional aspects as well as by identifying its model-based features – for example, those that are involved in ethical deliberation through imagery or emotions.

We know that almost all moral decisions are based on *incomplete information*. In *abductive reasoning*, when we base our hypotheses on incomplete information, we are forced to make *nonmonotonic* inferences, and we then draw defeasible conclusions from incomplete information.[11] Traditional deductive logics are always monotonic: intuitively, adding new premises (axioms) will never invalidate old conclusions. In a non-monotonic system, however, when the number of axioms, or premises, increases, the number of theorems does not. Also, because ethical deliberations always take place without complete information – even the best-informed people cannot know everything, after all – we must frequently abandon previously derived plausible/good hypotheses when we discover and add new data. This is almost always true in ethics, but it is particularly compelling in the case of utilitarian deliberations like the one undertaken by the parents of Baby Jane Doe.

Consequently, the data used in such deliberations must be assumed to be incomplete. This nonmonotonic feature was illustrated when Baby Jane Doe's unexpectedly good outcome reclassified the initial deliberation as wrong; the medical knowledge and data available at the time should not ultimately have been considered "good reasons" to reject surgery. But, unfortunately, those reasons were the best ones the parents had at their disposal when making the decision. This methodological situation clearly illustrates the role of data and information, which were necessarily limited and incomplete. Every possible ethical deliberation gives rise to some expected consequences that must be considered: for example, utilitarian deliberations involve a process of anticipating and envisioning consequences that can be used systematically to check whether or not the deliberation itself is correct. Sometimes the consequences confirm the soundness of the deliberation; in other cases, they show the deliberation was wrong.

The Baby Jane Doe case also underscores the fact that when seeking good reasons to support a moral decision, it is possible (and sometimes necessary) for us to change our minds when new or unexpected information arises, and as a result we may reach concrete decisions that directly oppose our initial choices – and this can occur not only when utilitarian

[11] A rigorous definition of monotonic and nonmonotonic logical systems is given in chapter 7, in the section "Abductive Reasoning and Epistemic Mediators."

reason are applied. Keeping in mind pragmatic and temporal constraints, we must always identify as many principles as possible when evaluating a moral situation. Then, if new information renders a previously adopted reason inadequate, having already identified multiple possible alternatives makes it easier to replace the now-obsolete reason with a more effective one.

The reasoning involved in ethics is notably similar to the reasoning used in diagnosis, even if in the first case we are evaluating the legitimacy (and justification) of certain moral judgments and deliberations, and in the second case assessing the explanation of data:[12] in both instances, we rely on an existing set of possible answers. But when we bypass traditional, established options and seek new solutions, a different process is at work. Unlike making a diagnosis, identifying a new disease and understanding its causes requires *creative* reasoning; those processes are analogous to the discovery, for instance, of a new ethical principle or prototype. Such discoveries are different from medical diagnosis, where, instead of creating something new, the task is to *select* from an encyclopedia of pre-stored diagnostic hypotheses. All we can expect of diagnosis is that it tends to select for further examination hypotheses that have a chance of being the best explanation of a particular case. This selective reasoning produces diagnostic hypotheses that give at least a partial explanation of an illness and will, therefore, always have at least some initial plausibility. Thus, diagnosis involves selecting plausible diagnostic hypotheses, which is followed by deduction to explore their consequences and induction to test them against available data. These final two steps increase the likelihood that a hypothesis is correct by identifying evidence that supports it rather than competing hypotheses or, conversely, by noting evidence that refutes all but one hypothesis.

If new information does in fact appear during this first cycle of deliberation, other hypotheses become eligible for consideration, and a new cycle begins. In this case, the *nonmonotonic* character of diagnostic reasoning is clear. The same pattern occurs when interrogating a situation in ethics, which involves selecting an ethical principle or prototype and then withdrawing it, if necessary, when new data emerges and reveals that a different principle or prototype is more appropriate. In some cases, as I have shown in chapter 1, there are ethical situations in which it is difficult to anticipate the *consequences* of possible deliberations. Environmental problems are "consubstantial" with the human species, and the difficulty of predicting the ecological consequences of our choices arises from a chronic lack of philosophical, scientific, and moral knowledge. This situation is the legacy of our cultural and political traditions: we are

[12] I will furnish more details about this analogy and about the abductive structure of diagnosis in chapter 7, where I will also deal with the tradition of "casuistry" in ethics.

hindered by incomplete basic information, a lack of scientific and moral knowledge needed to create good predictive models, and inconsistent (or nonexistent) communication and democratic and participatory processes. These methodological shortcomings explain why the consequences of human actions – in ecological settings, for example – can be so unpredictable and why "today's solutions can all too easily become tomorrow's problems."[13]

Sometimes our actions bring unforeseeable results because of interdependencies in opaque networks and other concurrent behaviors, as in the case of actions we perform on the internet. We are, so to say, "epistemically" vulnerable; epistemology requires ethics, insofar as the correct use of available moral and scientific knowledge can be achieved only when certain kinds of morally adequate relationships exist between people.[14]

MODEL-BASED MORAL REASONING

We usually conceive of ethical knowledge and reasoning as expressed in words at a verbal/propositional level, as in the case of the well-known ethical principle "Do not kill." But they can also exist in nonverbal, model-based forms, and in these cases, manipulative/"through doing" aspects become important – as with, say, visual imagination and direct action, respectively. Moreover, moral thinking can be verbal/propositional (sentential), model-based, or a hybrid mixture of several aspects; if our goals are new moral knowledge and new perspectives, it can also be creative. Did not the novel strategy of civil disobedience give new value to certain human behaviors? When Gandhi urged fellow Indians to forgo British fabrics and weave their own cloth, for example, he imbued textile making with a new, nationalistic meaning. Likewise, the Montgomery, Alabama, bus boycott organized by Martin Luther King, Jr., turned the simple act of walking into a moral (and, of course, political) statement.

Model-Based Reasoning

One model-based way to approach moral problems is to envision an imaginary world and then to identify possible outcomes of various choices there rather than in real, higher-risk settings. Strategies for trying out

[13] Cf. Kirkman, 2002, p. 38. In chapter 4 I have illustrated the importance of knowledge for improving the possibility of predicting natural events and the consequences of human actions.

[14] It is interesting to note the recent flourishing of research in the area of uncertainty reasoning, from artificial intelligence to economics, from decision making to ecology, and so on (see, for example, Parsons, 2001).

and assessing a proposed action can also be adopted in a model-based way, and once imagined, this world can help us to anticipate any possible internal contradictions. I will consider the problem of moral deliberation through the use of "models" in this section.

The analysis of model-based conceptual change in science has helped researchers to study revolutionary transformations: different varieties of what I call *model-based* abduction[15] are related to some types of such conceptual change. Charles Sanders Peirce stated that all thinking is in signs, and that signs can be icons, indices, or symbols:

> It is sufficient to say that there is no element whatever of man's consciousness which has not something corresponding to it in the word; and the reason is obvious. It is that the word or sign which man uses *is* the man himself. For, as the fact that every thought is a sign, taken in conjunction with the fact that life is a train of thoughts, proves that man is a sign; so, that every thought is an *external sign*, proves that man is an external sign.[16]

It is by way of signs that we ourselves *are* semiotic processes – for example, we are a more or less coherent cluster of narratives that are, essentially, types of signs. If all thinking is in signs, then it is not true that thoughts are in us, because we are in thoughts. Moreover, *inference* is a form of sign activity, but we have to remember that for Peirce the word *sign* includes "feeling, image, conception, and other representation."[17] Put another way, a significant part of thinking activity is *model-based*. As we will see in the last part of the next chapter, scientific model-based reasoning exhibits the maximal cognitive relevance when exploited in the creative processes of scientific discovery. Model-based reasoning plays an important role in moral deliberation, as I will illustrate, but for now I would like to address the problem in general philosophical and cognitive terms. It is well known that Peirce ascribes great importance to diagrammatic thinking, a value manifested in his discovery of the powerful predicate logic system based on diagrams or "existential graphs": he considers not just conscious abstract thought, but any cognitive activity whatsoever – including perceptual knowledge and subconscious cognitive activity – to be inferential.[18] In subconscious mental activities, for example, visual representations – which are typical model-based devices – would play an immediate role, according to Peirce.

Consider how so-called visual reasoning begets a kind of model-based performance when we instantly form "hypotheses" about a given situation

[15] Magnani, 1999.
[16] Peirce (*CP*), 5.314.
[17] Ibid., 5.283.
[18] Davis, 1972.

using information from similar previous experiences. Such thinking features a mental process that falls into the category of perception. Philosophically, Peirce views *perception* as a fast and uncontrolled knowledge-production procedure. Knowledge constructions are instantly reorganized by perception, and consequently, they become habitual and diffuse and do not need any further testing: "… a fully accepted, simple, and interesting inference tends to obliterate all recognition of the uninteresting and complex premises from which it was derived."[19] Many visual stimuli – which can be considered the "premises" of the involved abduction – are ambiguous, yet people are adept at imposing order on them: "We readily form such hypotheses as that an obscurely seen face belongs to a friend of ours, because we can thereby explain what has been observed."[20] I have called this kind of inference *visual abduction*, a special kind of nonverbal abduction that allows us to form image-based hypotheses.[21]

Of course, visual model-based reasoning (conscious or subconscious) in everyday cognitive behavior is not of particular interest here, but in science it may be very significant and lead to interesting new discoveries.[22] Moreover, perceptions, like scientific hypotheses, may be withdrawn; in a way, they are "hypotheses" about data that we may accept – sometimes spontaneously – or carefully evaluate. In all these examples, Peirce is referring to a kind of hypothetical activity that is inferential but not verbal, where "models" of feeling, seeing, or hearing, and so on effectively build both the habitual abductions of everyday reasoning and the creative abductions of intellectual and scientific life.

Following Nancy Nersessian,[23] I use the term "model-based reasoning" to indicate the construction and manipulation of various kinds of representations – not just those that are sentential and/or formal. There are various common forms of model-based reasoning: for example, constructing and manipulating visual representations, thought experiments, analogical reasoning, and the so-called tunnel effect,[24] which occurs when models are built at the intersection of some operational interpretation domain and a new, poorly understood domain. We have to remember that visual and analogical reasoning can be usefully employed to form scientific concepts; such ideas do not simply emerge fully formed from a scientist's head, but rather evolve during problem-solving activity that involves various model-based procedures: this process is a *reasoned*

[19] Peirce (*CP*), 7.37.
[20] Thagard, 1988
[21] Cf. Magnani, 2001a, pp. 33–49.
[22] Magnani et al., 1994; Shelley, 1996.
[23] Nersessian, 1995a and 1995b.
[24] Cornuéjols et al., 2000.

process. It is also important to remember that reasoning usually occurs in a multimodal way, in the sense that its sentential and model-based components can be more or less strictly intertwined and distributed.

Finally, we must note that our characterization of model-based reasoning is mainly cognitive and refers to forms of reasoning that exploit models like visualizations, various simulations, analogies, and so on. A reader might reasonably wonder about the relationship between these systems and current logic: from the logical point of view, recently built systems provide "deductive" models of model-based reasoning. Some of these systems are called "heterogeneous": within a demonstrative framework, they produce representations that derive from a variety of representational systems that are sentential, model-based, diagrammatic, and usually considered nondemonstrative. Their advantage is that they allow a reasoner to bridge the gaps between various formalisms and to construct threads of proof that cross the boundaries of the systems of representation.[25]

Moreover, because model-based reasoning is particularly well suited for creative endeavors, it shares with counterfactual thinking (and related counterfactual logical systems) a disposition for constructing imagined interventions and for reasoning causally about possible worlds and how things might otherwise be. It must be said that in general, counterfactual logics have considered only sentential aspects, unlike the heterogeneous systems.[26]

What is the role of model-based reasoning in ethics and moral deliberation? Mark Johnson, for example, considers this kind of reasoning to be fundamental to ethics: "Moral principles without moral imagination become trivial, impossible to apply, and even a hindrance to morally constructive action."[27] This means that analogical and metaphorical model-based reasoning conducted in the imagination is very important because of its capacity to "reconceptualize" the situation at hand. Consequently, model-based tools for ethical deliberation should not necessarily be rejected as "subjective, free flowing, creative processes

[25] Swoboda and Allwein, 2002. In doing this, heterogeneous systems allow the reasoner to take advantage of each component system's strengths. For example, "recast" rules are clearly elicited as rules of inference that allow information exchanges among the various representations. Such rules allow us to extract data in two ways: they help us to derive from diagrams information to be expressed in a sentential form, and they allow us to glean from a formula information to be incorporated in a diagram. Clarifications of the exact processes and semantic requirements of manipulative inferences and distributed cognition are given in Shimojima, 2002.

[26] A recent paper that connects counterfactual reasoning to cognitive aspects is Sloman and Lagnado, 2005.

[27] Johnson, 1993, p. x.

not governed by any rule or constrained by any rationally defined concept" so that imagination appears to be an "enemy of morality."[28]

Schematism and Typification

Immanuel Kant was perfectly aware that intermediate thinking devices allow human beings to apply abstract principles to the world of experience. In the *Critique of Pure Reason*, Kant specifically studies the example of the geometrical construction, which is what we produce when we sketch a geometrical figure in hopes of discovering inferences that will lead to the right solution. Kant says in the "Transcendental Doctrine of Method":

> To *construct* a concept means to exhibit *a priori* the intuition which corresponds to the concept. ... Thus I construct a triangle, by representing the object which corresponds to this concept, either by imagination alone, in pure intuition, or in accordance therewith also on paper, in empirical intuition – in both cases completely *a priori*, without having borrowed the pattern from any experience.[29]

More precisely, there is a universal schematic activity driven by the imagination – Kant sometimes classifies it as a *rule*, sometimes as a *model*, and on other occasions as a *procedure* – that facilitates the transformation of pure geometrical concepts into sensible representations.[30] We can say that the activity of schematism is implicit, inscrutable, arising mysteriously from the imagination in a way that is impossible to analyze rationally:

> This schematism of our understanding, in its application to appearances and their mere form, is an art concealed in the depths of the human soul, whose real modes of activity nature is hardly likely ever to allow us to discover....[31]

And then,

> ... the result of the power of imagination [is] ... a blind but indispensable function of the soul, without which we should have no knowledge whatsoever, but of which we are scarcely ever conscious.[32]

This example explains how a pure intellectual (geometrical) concept can be applied to experience. But what happens in the case of maxims or other abstract moral rules? In ethical deliberation, there is no direct

[28] Ibid., p. 2.
[29] Kant, 1929, A713/B741, p. 577.
[30] On the role of schematism in the Kantian philosophy and philosophy of geometry, see Magnani, 2001b.
[31] Kant, 1929, A141/B181, p. 183.
[32] Ibid., A78/B103, p. 112.

schematism; the laws of freedom (ethical rules) cannot be directly represented in the concrete world, which operates according to the laws of physical necessity, not those of freedom. As Kant says,

> A schema is a universal procedure of the imagination in presenting a priori to the sense a pure concept of the understanding which is determined by the law; and a schema must correspond to natural laws to which objects of sensuous intuition as such are subject. But to the law of freedom (which is a causality not sensuously conditioned), and consequently to the concept of the absolutely good, no intuition and hence no schema can be supplied for the purpose of applying it *in concreto*. Thus the moral law has no other cognitive faculty to mediate its application to objects of nature than the understanding (not the imagination); and the understanding can supply to an idea of reason not a schema of sensibility but a law. This law, as one which can be exhibited *in concreto* in objects of the senses, is a natural law. But this natural law can, for the purpose of the judgment, be based only in a formal aspect, and it may, therefore, be called the *type* of the moral law.[33]

Interpreting this passage is difficult, but it relates to the idea that a pure moral rule (as a maxim of action) is applied to a concrete experience as a kind of "typification" – a sort of figurative substitute (which could be the analogue of the schematism for pure concepts, like geometrical ones, and for empirical concepts): this typification could be interpreted as a kind of figurative envisioning of a nonexistent world as a means for judging a given moral situation. But Kant, denying that this typification involves imagination, maintains that moral judgment is a matter of pure practical reason.

Nevertheless, I agree with Johnson when he concludes, "... what could be more thoroughly imaginative than this form of figurative envisioning that is based on a metaphoric mapping?"[34] Anyway, it is in this way, Kant says – through typification – that we can treat a moral law as if it were a law of nature.

Moral Model-Based Reasoning

Going beyond rules and principles, we know that prototypes, schemas, frames, metaphors, and narratives are all vehicles for model-based moral knowledge and can be very effective ways to deal with moral problems. A typical moral prototype, for example, could be the "morality" of grammar. Grammatical principles can be seen as parallel to moral principles, as in the simple case of the analogy between "speaking well" and "behaving well." Just as prescriptive grammar dictates that, for example, "I have *fewer* worries" is proper usage, but "I have *less* worries" is not, so

[33] Kant, 1956, p. 69.
[34] Johnson, 1993, p. 73.

too does a moral principle privilege certain behaviors as the "correct" ones.

Another example is related to the idea of action as a metaphorical "motion," which leads to the idea that moral principles can be considered rules telling us which "action paths" we may take, which ones we must take, and which we must never take.[35] Often in the case of schemas, prototypes, and metaphors (as in this case of the "grammar" metaphor just mentioned) the model-based level (visualizations, simulations, feelings, etc.) is accompanied by the propositional level (in terms of sentences and abstract verbal structures), as we will see immediately in the case of the morality of marriage.

This pairing of the model-based with the propositional occurs in the case of the morality of marriage. Johnson offers many interesting metaphors/prototypes for the concept: he compares it to a manufactured object, an ongoing journey, a durable bond between two people, an investment, an organic unity, a resource, and a commodity exchange. These systematic metaphors lead to organized mappings of objects, events, causes, and relationships from one domain onto the other. Alex, a character in one of Johnson's examples, thinks of his marriage as a resource: it "seemed important then," he muses, "to have someone there all the time that you could rely on. And talk to all the time about things. Somebody to help and somebody to help you, that seemed like a good idea," and it is this metaphor that informs him as he chooses ethical directions in managing the marriage. Moreover, if we cast marriage as an organic unity, the following moral duties immediately derive: (1) a demand for monogamy, a special exclusive relationship between husband and wife; (2) the requirement for physical closeness or proximity in order to preserve that unity; and (3) the obligation to share experience (mutual growth).[36]

Moral reasoning narratives, which, of course, are clearly propositional/verbal rather than model-based, sometimes also include model-evoking structures. Human beings are ideal hardware for narrative, especially from the ethical point of view: we are storytelling animals. There are prototypical narratives in the form of "linguistic stories we tell to others and sometimes write down"[37] that constitute spoken and written texts concerning moral "exemplars." We usually see our own lives and those of others as a series of narratives, and we "continually reinterpret and revise our narrative self-understanding." For example, we scroll through our cache of stories to find one that can best clarify the moral problem at

35 Ibid., p. 43.
36 Ibid., pp. 53–54.
37 Ibid., p. 152.

hand and that we can reconcile with our self-representations and ideals.[38] Narratives of this sort can evoke not only images and analogies but also emotions and feelings (cf. the following section).

These considerations remind us that, as we have many times illustrated in the previous chapters, we need to regard people as both ends and "things"; this dual respect will allow us to enrich others with new values and enhance our ethical understanding of their problems. Historically, when we speak of a manufactured object, a commodity exchange, an organic unity, an ongoing journey, and so on, we have in mind a nonhuman thing, but these terms can also refer to human relationships – as in our example of marriage – and seeing relationships in this way can help us to reshape them.

Finally, to continue our discussion about using an imagined setting to predict the possible consequences of our choices, we must stress that this strategy can be enacted in a model-based way – using a figurative approach, for example, as demonstrated in the previous section. Once we have "constructed" the stage for our experiment, we must then determine whether a proposed solution would create some sort of internal contradiction in this nonexistent world. In his discussion of a model-based environmental ethical problem, Bryan Norton describes the so-called demand models as examples of a "mission-oriented" science, understood as communicative model-based devices able to help in solving local ecological problems in the event of failures of communication that inhibit policy development. The case study is based on research conducted on the management of Lake Lanier in northeastern Georgia. Competing interest groups employed different mental models to find solutions to the lake's ecological problems, taking advantage of a very interesting "demand model" and exploiting some sort of analogical reasoning. They found that it is possible to improve communication and cooperation on ecological problems and policies by creating a shared and unified framework of the lake that clarifies the differences in those mental models.[39]

Moral Emotions

I have illustrated that Peirce categorizes all knowing as a form of *inferring*, and as we know, inferring is not instantaneous – it is a process that requires that many kinds of models be used over a fairly considerable period of time. Hence, all model-based reasoning is a kind of inference, which does not necessarily contradict his notion that the *inferential* character of reasoning is based on instinct (the mind is "in tune with

[38] Ibid., pp. 152–153.
[39] Norton, 2002.

nature"). To follow Peirce further, all sensations or perceptions also contribute to a unifying hypothesis in the case of emotions:[40]

> Thus the various sounds made by the instruments of the orchestra strike upon the ear, and the result is a peculiar musical emotion, quite distinct from the sounds themselves. This emotion is essentially the same thing as a hypothetic inference, and every hypothetic inference involved the formation of such an emotion.[41]

In a Peircian sense, emotions also have an "inferential" character and employ a kind of model-based process. Human beings and animals have evolved so that they can quickly recognize and respond to recurring events; emotions trigger an instant reaction to, say, a possible threat, so that the animal does not need to reassess a dangerous situation every time it arises: something like the case of a mouse that fears cats and has a better chance of survival than a mouse that doesn't. During evolution, such hypothetical types of recognition and explanation settled into some animals' nervous systems.[42] We can "hypothesize" fear as a reaction to a possible external danger; in people, such reactions can occur even when a person knows the threat is not real – when watching a suspenseful movie or reading a thriller.[43]

In decision making, emotions play a pivotal role: they speed up the process and lead directly to action. But using them to make choices is usually considered irrational because of their disadvantages: in the throes of strong feeling, we may be blind to some options, overlook critical information, or, when participating in a group charged with making a collective decision, fail to engage or connect with others who do not share our emotional state.[44] I think it is important to understand, however, that emotions are not inherently irrational. In fact, they can be useful tools in moral decision making if they are successfully intertwined with learned cultural behaviors so that they become "intelligent emotions." Emotions can be developed, and, as Picard points out, "Adult emotional intelligence consists of the abilities to recognize, express, and have emotions, coupled with the ability to regulate these emotions, harness them for constructive purposes, and skillfully handle the emotions of others."[45] There is an ongoing debate about the use of the expression "emotional

[40] Considering emotions as abductions, Oatley and Johnson-Laird (1987) have proposed a cognitive theory of emotions based largely on Peircian intuitions.

[41] Peirce (*CP*), 2.643.

[42] That is, in human beings, emotional abductive skills; cf. Magnani, 2001a, p. 45.

[43] Oatley, 1996.

[44] Thagard, 2001, pp. 356–357.

[45] Picard, 1997, p. 49. See also Nussbaum, 2001.

intelligence": while the word *intelligence* implies something innate, many aspects of emotional intelligence are actually skills that can be learned.[46]

Antonio Damasio differentiates conscious "feeling" from unconscious "emotion" (of course, only conscious individuals "feel"):[47] the genesis of emotions also relates to an individual animal's need to respond with its whole body – to run away from danger, for example, or to care for offspring. Emotions can communicate information about a situation and trigger a response even in the absence of consciousness; in turn, this holistic response seems to have influenced the evolutionary formation of self and consciousness.[48]

Happiness, sadness, fear, anger, disgust, and surprise all can be viewed as *judgments* about a person's general state; a man who unexpectedly comes across a tiger on the loose, for example, would be understandably afraid because the large carnivore threatens his instinct to stay alive. In fact, all emotions are connected to goal accomplishment: people become angry when they are thwarted, for instance, and feel pleased when they are successful. In this sense, emotion is a *summary appraisal* of a problem-solving situation. Moreover, it provides cognitive "focalization" of the situation and readies us for action. Alternatively, we can consider emotions as *physiological reactions* rather than as cognitive judgments. Damasio refers to the signals that the body sends to the brain as *somatic markers*. Neuroscience has taught us that emotions depend on the interaction between bodily signals and cognitive appraisal. That is, they involve *both* judgments about how the current situation is affecting our goals and neurological assessment of our body's reaction to that situation.

Emotions are represented in the brain, but they cannot be represented as concepts because this conceals their links to judgment, physiology, and feeling. We can imagine emotions as patterns of activation across many neurons in many brain areas, including those sites involved in cognitive judgment, like the prefrontal cortex, and those that receive input from bodily states, like the amygdala. Put another way, emotion activates neurons in different areas of the brain, areas that may have either inferential or sensory functions.[49]

We have already pointed out that emotions play an important role in moral deliberation. Damasio hypothesizes that in some cases of brain damage to the ventromedial, bottom-middle, or prefrontal cortex areas,

46 On the neurological and cognitive role of emotions, see Damasio, 1994 and 1999. A cognitive theory of moral emotions in terms of "coherence" is illustrated in Thagard, 2000 and 2001. For recent work on emotional intelligence, see Ben-Ze'ev, 2000; Matthews, Zeidner, and Roberts, 2002; and Moore and Oaksford, 2002.

47 Damasio, 1999.

48 Modell, 2003. On consciousness, conscious will, and free will, see the chapter 3 section "Critical Links: Consciousness, Free Will, and Knowledge in the Hybrid Human."

49 Wagar and Thagard, 2004.

people lose the ability to make effective decisions because they cannot discern the future consequences of their actions, especially in social contexts. That part of the brain provides connections between areas of the cortex involved in judgment and areas involved in emotion and memory, the amygdala and hippocampus.

Computer simulations of decision making have made it clear that we need more neurologically "realistic" models involving the role of emotions.[50] The GAGE neurocomputational program[51] aims at filling this gap. It models the cognitive situation, just described, which is caused by a brain lesion, using groups of computational spiking neurons corresponding to each of the crucial brain areas involved: (1) the vetromedial frontal cortex, (2) the amygdala, and (3) the nucleus accumbens (a region strongly associated with rewards). So GAGE is capable of taking into account both cognitive aspects of judgment and appraisal performed by the ventromedial prefrontal cortex and physiological input mediated by the amygdala.[52]

Let us revisit Johnson's take on moral problems related to marriage as an organic unity as he describes the roles of emotions, feelings, desires, and motivations for a husband named Alex:

> They involve entertaining in imagination the presence of one's absent spouse, the way this absence makes shared experience impossible, and the memory of earlier shared moments. *There is no separation of intellectual precepts from feeling and imagination*. Alex is motivated, just as all morally sensitive people are, not by the alleged purity of rules and abstractions, but rather by feeling and imagination that draw him to desire something as good. About this Hume was quite right, stressing as he did that an emotional dimension is crucial in our moral deliberations. He was wrong, unfortunately, to separate this emotional aspect from what he mistakenly took to be a pure cognitive or intellectual component. In fact, these two dimensions are blended in a way that makes it impossible to extract one from the other. One reason metaphor is so basic for our moral understanding is that it combines these very dimensions of our embodied moral awareness – projection of possibilities, relation of feelings, imaginative reflection – that make it possible for us to have any degree of moral sensitivity in the first place.[53]

I think this interesting passage suggests that feelings serve as moral inferences especially if they are intertwined with learned cultural behaviors

[50] Of course, they will still lack the possibility of "feeling" emotions: indeed, they will not have "bodily" inputs.

[51] Ibid.

[52] Some meta-ethicists call moral intuitionism the view of emotions as central in justifying moral beliefs (Sinnott-Armstrong, 1996). Ben-Ze'ev (2000) maintains that optimal moral behavior is that which combines emotions and intellectual reasoning, a complex integration that requires the so-called emotional intelligence.

[53] Johnson, 1993, pp. 58–59.

and therefore become "intelligent emotions" or, as some ethicists say, "appropriate" emotions.[54] Hume was wrong to view emotions as separate from intellect, says Johnson, who maintains that the two dimensions are actually blended. I do not think we need to blend rational and emotional aspects; emotions are still emotions, even though we may be tempted to lump them together with purely cognitive functions because they can function "morally" under certain conditions – that is, when they are not just raw products of evolution but are, instead, further shaped by knowledge and information.

Hence it can be said that sensations of pain, pleasure, harm, and well-being can arise from instinct in some instances and from moral training and experience in others. The latter scenario arises when, for example, we acquire a virtue like kindness that then automatically informs our behavior; the resulting spontaneous acts of kindness subsequently generate positive feelings not only for those we help but also – and more to the point here – for ourselves.[55] Such behavioral habits are activated when we confront moral problems, and it is in this constrained and educated way that love and compassion and goodwill can guide us "morally."[56] The strong feelings we may experience about a particular issue are an indication of moral depth and empathy and are, therefore, the manifestations of a precious aspect of human nature. Nevertheless, as I have already said, emotions can impede good moral decision making if they result from irrational prejudices, selfishness, or cultural conditioning, all of which make it difficult to consider good arguments for opposing moral views. In such cases, one may be inclined to engage in the "nasal reasoning and olfactory moral philosophy.[57]

I agree wholeheartedly with Peirce that all thinking is in signs and that these signs take many forms – icons, indices, symbols, and so on, as mentioned earlier. If all *inference* is, in fact, a form of sign activity – as Peirce contends – and if we use the word *sign* to include feelings, images, conceptions, and other representations, then we must include unconscious thought among the model-based ways of moral thinking. Indeed, it

[54] Thomson, 1999, pp. 148–149. In this case, Oatley (1992), speaks of "learned spontaneity."

[55] Cf. Evans, 2001. The interesting role of virtue ethics in the health professions is illustrated in Pellegrino, 1999.

[56] Joyce (2006) also thinks that language is a prerequisite for having certain educated moral emotions, especially guilt and its ilk. Emotions like guilt, shame, and disgust – which are not present in infants or animals – are "cognitively rich," he says, and like other concepts necessary for particular emotions, they are evaluative concepts. "[L]anguage is necessary for certain emotions and ... the evolution of language made certain emotions accessible" (p. 94).

[57] Cf. our chapter 2 discussion of visceral rather than intellectual reactions to cloning. Other examples of irrational "olfactory and nasal philosophy" that function against gay people are given in Mohr, 1998. See also Leiser, 2003.

is not just conscious abstract thought that we can consider inferential: we can characterize many cognitive activities in this way. Consider our example of Alex and his attitude toward his marriage – it can be hypothesized that his strong commitment to his wife relates to his understanding of marriage, which in his case is a stable, well-formed, tacit, unreflective understanding. The same forces are at work when we unconsciously reject the moral legitimacy of assisted suicide because of a spontaneous, emotional, often implicit personal fear of death.

Martha Nussbaum has emphasized the cognitive value of emotions and further clarified their moral role; in her work, she improves and updates the Greek Stoic view, which holds that emotions are evaluative judgments that ascribe great importance to certain things and persons outside a person's own control. From this perspective, such things then have the power to determine human flourishing (*eudaimonia*).[58] Put another way, through emotions, people acknowledge that external things/persons they do not fully control are very important for their own flourishing. Emotions are always seen as involving thought of an "intentional" object combined with thought of that object's salience, value, and importance in the framework of what Nussbaum calls the "cognitive-evaluative" view. This perspective contrasts with the "adversary view," which considers emotions mere "nonreasoning movements" that derive "bodily" from an animal part of our nature rather than "mentally" from a specifically human part. It is difficult to see emotions as judgments in this view, as opposed to the stoic view, and it would seem hard "to account for their urgency and heat given the facts that thoughts/judgments are usually imagined as detached and calm":[59] emotions are fundamentally seen as irrational and as a bad guide to action in general and to moral action in particular.

From the "cognitive-evaluative" perspective, it is very easy to think of emotions more broadly in a way that goes beyond the Stoic starting point; consider, for example, the cognitive role of emotions in animals and the evaluative appraisal they perform in moral life. Of particular interest is the role that elements of culture – social norms, for example – play in shaping certain feelings like compassion that I refer to as "trained emotions":"Human deliberative sociability also affects the range of emotions of which humans are capable,"[60] just as individual history influences the perception of that effect and cognitively embeds emotions

[58] Greek eudaimonistic ethical theories are concerned with human flourishing: *eudaimonia* is taken to include everything to which a person attributes intrinsic value. Nussbaum retains this spelling, rather than using the English word "eudaemonistic," because she wants to refer to the ancient Greek concept and avoid the more recent connotations of the idea, "namely, the view that the supreme good is happiness or pleasure" (2001, p. 31).

[59] Ibid., p. 27.

[60] Ibid., p. 148.

in a complex of personal narratives. In this last sense, to acknowledge the influence that social constructions have on emotions is to see that emotions consist of elements we have not ourselves constructed.

Finally, I must note that Michael Gazzaniga, citing James Q. Wilson's prediction, provides a brain-based account of moral reasoning centered in the areas of emotion. It has been found that brain regions normally involved in emotional processing may be activated by one type of moral judgment but not by another. In this sense, when someone is willing to act on a moral belief, it is because the emotional side of her brain was activated as she checked the moral question at hand. If, on the other hand, she decides not to act, the emotional part of the brain has not become active.[61]

In discussing some of Paul MacLean results,[62] David Loye illustrates the possible evolutionary origins of moral capacities and moral agency, developed in the so-called theory of GSHM (Guidance System of Higher Mind).[63] This is a general model of intelligence in which moral functions are integrated with cognitive, affective, and conative functions, resulting in a flow of information among eight brain levels working as an evaluative unit between stimulus and response. This model is based on Charles Darwin's theory of the grounding of morality in sexual instincts that later expand into parental love; the model also takes advantage of research on animals. The study shows how prefrontal function is closely related to the development of moral sensibility and judgment (on the assumption that emotional responses are closely tied to reason through the functions of the prefrontal cortex).

BEING MORAL THROUGH DOING: TAKING CARE

In science, there exists what I have called manipulative abduction[64] – that is, discovering new ideas and theories by manipulating external objects and representations (also, of course, by exploiting the embodied and tacit skillful capacity to act). I contend that manipulative abduction is not limited to the scientific realm – it is also present in ethical reasoning.

[61] Gazzaniga, 2005. Other studies exploiting neuroimaging have dealt with the neuro-anatomy and neuro-organization of emotion, social cognition, and other neural processes related to moral judgment in normal adults and in adults who exhibit aberrant moral behavior (Greene and Haidt, 2002). Moll and Colleagues (2002, 2005) have established that moral emotions differ from basic emotions in that they are interpersonal: the neural correlates that are more involved in moral appraisal appear to be the orbital and medial sectors of the prefrontal cortex and the superior sulcus region, which are also critical regions for social behavior and perception.

[62] MacLean, 1990.

[63] Loye, 2002. On this issue cf. also de Waal, 2006.

[64] Cf. the following chapter, the section "Abductive Reasoning and Epistemic Mediators."

Manipulative and "through doing" aspects are important, as when morality results directly from action. A considerable part of morality is enacted in a tacit way, "through doing," as I say. Moreover, part of this "doing" can be seen as a manipulation of the external world as a way to establish "moral mediators" in order to achieve certain ethical effects. Nevertheless, we already know that moral mediators are also beings, entities, objects, structures that bring unintentional consequences – either ethical or unethical – as we have seen in the earlier chapters. Peirce gives another interesting example of model-based abduction that is related to sense activity: "A man can distinguish different textures of cloth by feeling: but not immediately, for he requires to move fingers over the cloth, which shows that he is obliged to compare sensations of one instant with those of another."[65] This surely suggests that reasoning and inferential processes also have interesting extratheoretical characteristics. Moral inferences also have a role in manipulating various external objects and nonhuman structures as substitutes for and supplements to moral "feeling" and "thinking": there is a morality through doing.[66]

In science, this kind of manipulative reasoning happens when we are thinking *through* doing in a pragmatic way, not just when we are thinking *about* doing; such reasoning uses external objects and representations in a way that goes beyond the well-known role of experiments. Thinking through doing allows us to use results – nature's answers to the investigator's question – to do more than just form new scientific laws or predict certain outcomes through confirmation or falsification.

This kind of extratheoretical cognitive behavior is also manifested in many everyday actions that people perform perfectly well without needing any conceptual explanation. Established "conceptual" accounts can deteriorate, and in such cases, procedural know-how that was once explicitly present in the memory must be reproduced. When first learning to drive a car with a manual transmission, for example, one must constantly be conceptually aware of the need to press on the clutch before shifting gears, and at that point the conceptual account is in the foreground of memory. Eventually, though, drivers learn to move through the gears without really being aware that they are doing so. If, however, a driver later buys a vehicle with an automatic transmission and drives it exclusively for many years, gear-shifting know-how recedes to the background of memory, and its vivid presence deteriorates. Then, if she is for some reason required to drive a car with a manual transmission again, the driver must actively retrieve the related know-how. Usually, she is able to

65 Peirce (*CP*), 5.221.

66 On the role of the "computational circuitry of human cognition" that "flows both within and beyond the space of human thought and reason," see Clark, 2002, p. 26; Clark, 1997; Dascal, 2002; and Mun Chan and Gorayska, 2002.

perform the task spontaneously, which means that her cognition is embodied and does not need the awareness typical of "conceptual" knowledge. Of course, this sort of process occurs only when dealing with established knowledge sets – in other situations, as we have said, an account must be constructed for the first time, as in the case of creative scientific manipulative abduction, where the manipulation – of an experimental artifact, for instance – is embodied and new.[67]

Another type of embodied cognition is illustrated by Edwin Hutchins in his description of a navigation instructor who, after automatically performing complicated plotting manipulations and procedures on a regular basis for three years, was suddenly struck with an insight about the conceptual relationship between relative and geographic motion – not when he was working, but "as he lay in his bunk one night." This example illustrates that learning often occurs when we impose new conceptual and theoretical details on manipulative executions that are so familiar that we normally conduct them without thinking. The instructor does not discover anything unknown about the involved skill, but we can say that the discovered knowledge is new, at least for him.[68]

Certainly much of the thinking agent's complex environment is internal; our individual knowledge bases work with our personal inferential and emotional expertise as a sort of software. Nevertheless, any cognitive system – whether in a person, an "external" object, or a technical artifact – features "distributed cognition."[69] Take, for example, the diagrams used in elementary geometrical reasoning. Specific figures serve as states, and the implied operators[70] are the manipulations and observations that transform one state into another. The geometrical outcome depends on specific sensorimotor activities performed on an object, which acts as a dedicated external representational medium supporting the various operators at work. There is a kind of an epistemic negotiation between the sensory framework of the geometer and the external reality of the diagram. This process involves an external representation consisting of written symbols and figures that are manipulated by hand. The cognitive system, then, involves more than just the mind-brain's performance of the geometrical task: it encompasses the geometer's whole body (cognition is *embodied*) in addition to the external physical representation. In geometrical discovery, then, the act of cognition is located – distributed – in a system consisting of a human being together with external diagrams. The type of external representation one

67 Cf. the following chapter, the section "Thinking through Doing: Manipulative Abduction."

68 Hutchins, 1995.

69 Ibid.; Norman, 1993.

70 Magnani, 2002a.

uses to solve a problem can have an effect on the computation itself: the Roman numeration system, by means of external signs, eliminates the more difficult aspects of addition, whereas the Arabic system simplifies parts of complex multiplications.[71] David Gooding refers to this kind of concrete manipulative reasoning in science when he describes the so-called construals that embody tacit inferences in visual and tactile performances that often involve apparatus and machines.[72]

Finally, as I have already said, it has to be noted that reasoning usually occurs in a hybrid way, in the sense that the various components (sentential, model-based, manipulative) are more or less interrelated and suitably distributed.

Templates of Moral Doing

As I have demonstrated in previous research on explanatory and creative processes in reasoning, "thinking through doing" and manipulative abduction are two central aspects of scientific activity. I would now add that a considerable part of morality is also achieved in a similar tacit and manipulative way; it is, so to speak, a morality "through doing." Moreover – and this is another crucial idea of this book – part of this "doing" can occur through manipulating the external world in order to build various distributed kinds of "moral mediators."

It is impossible to compile an exhaustive list of invariant behaviors that can be considered basic forms of ethical manipulative reasoning. Generating moral effects through the use of nonhuman objects in real or artificial environments requires that old and new *templates* of behavior be repeated at least somewhat regularly. Only rarely are we referring here to action that simply follows "explicit," articulated, previously established plans; at issue are embodied, implicit patterns of behavior that I call tacit templates. This type of "hidden" moral activity is still conjectural: these templates are embedded moral hypotheses that inform both new and routine behaviors and that, as such, enable a kind of moral "doing." In some situations, templates of action can be *selected* from those already stored in the mind-body system, as when a young boy notices his baby sister crying and, without thinking, automatically tries to comfort the infant by stroking her head or singing a lullaby, as he has seen his parents do many times. In other instances, however, when an agent has no established protocol to draw on, new templates must be *created* in order to achieve certain moral outcomes. Such newly forged behavior patterns are, as we will see, important components of the concept of knowledge as duty.

71 Zhang and Norman, 1994; Zhang, 1997.

72 Gooding, 1990. For further details, see the chapter 7 section "Thinking through Doing: Manipulative Abduction."

New challenges require new templates, and in this book we have illustrated many new challenges, like those generated by technological products.

The following tacit templates of moral behavior (see Figure 6.1)[73] feature some interesting characteristics:

1. Sensitivity to *curious or anomalous aspects* of the moral situation. In this case, manipulations are performed in order to reveal potential inconsistencies in received knowledge, as when we suddenly adopt a different attitude toward our spouse in order to elicit a reaction that confirms or discounts hypotheses about his or her feelings, or in order to develop new hypotheses about the relationship. This might be the case when one spouse, assuming that tolerance is a measure of love and devotion, behaves in an annoying way in order to test the other spouse's level of acceptance. Another example could involve detectives, who, while investigating a crime, automatically analyze existing evidence in order to extract additional information or spontaneously look for further data about a suspect.
2. Preliminary sensitivity to a moral situation's *dynamical character* – not just to its entities and their properties. A common aim of manipulations is to practically reorder the dynamic sequence of the events and human relationships associated with the moral problem in order to find new options for action. An example might be a woman who, having already decided to have an abortion, unintentionally modifies her actions in a way that allows her to envision not going through with the procedure. She is unconsciously changing her behavior in hopes of making herself decide against the abortion.
3. Referral to manipulations that exploit *artificially created environments* and *externally induced feelings* to uncover stable, reproducible information about moral knowledge and constraints. This template is at work, for example, when a collectivity faced with sentencing a convicted murderer exploits resources like statistics, scientific research, and information from interviews in making a decision. These strategies produce real data rather than the sort of faulty information that would be derived from other sources – say, for example, the genuine relief felt by the murder victim's relatives when the criminal is killed. In this way, some aspects of the social orders of the affected groups are reconfigured.[74]

73 In this and the following figure, arrows represent a kind-relation.

74 On the reconfiguration of social orders that is realized in science (laboratories), see Knorr Cetina, 1999.

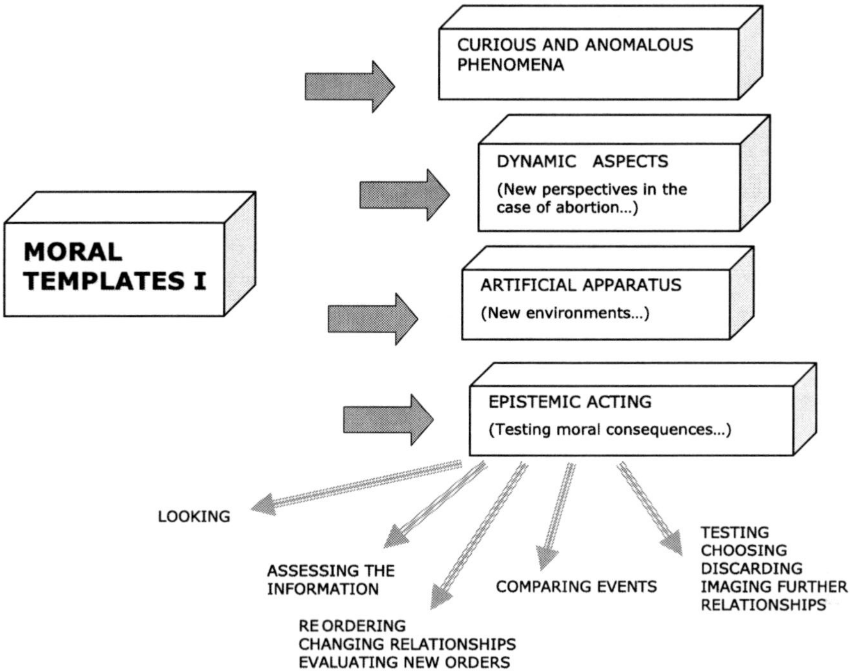

FIGURE 6.1. Conjectural moral templates I.

4. Various contingent ways of spontaneous moral acting. This fourth characteristic involves a cluster of very common moral templates. A person will automatically look at issues from different perspectives; *assess* available information; *compare* events; *test*, *choose*, *discard*, and *imagine* additional manipulations; and implicitly *evaluate* possible *new orders* and *relationships* (or simplify them in order to make new analogies or comparisons possible). All these strategies produce useful evidence that can be used to test previously established moral judgments and, in some cases, to improve the chances of accurately anticipating the nature and range of those judgments' consequences.[75]

Additional tacit template features relate to the following issues:

5. Spontaneous moral action that can be useful in the presence of *incomplete or inconsistent information* or a *diminished capacity to act morally* upon the world. Such action operates on more than just a

[75] Analogues of all these manipulative templates are active in epistemic settings: see Magnani, 2001a; Magnani, Piazza, and Dossena, 2002.

"perceptual" level – it is also used to get additional data that restores coherence and/or improves deficient knowledge.

6. *Action as a control of sense data* illustrates how we can change the position of our bodies (and/or of external objects) to reconfigure social orders and collective relationships; it also shows how to exploit artificially created events in order to get various new kinds of stimulation. Action of this kind provides otherwise unavailable tactile, visual, kinesthetic, sentimental, emotional, and bodily information that, for example, helps us to take care of other people (cf. the following subsection).
7. Action enables us to build new *external artifactual models* of ethical mechanisms and structures (through "institutions," for example) to substitute for the corresponding "real" and "natural" ones. (Keep in mind, of course, that these "real" and "natural" structures are also artificial – our cultural concept of "family" is not a natural institution.) For instance, we can replace the "natural" structure "family" with an environment better suited to an agent's moral needs, which occurs when, say, we remove a child from the care of abusive family members. In such a case, we are exploiting the power of an *artificial* "house" to reconfigure human relationships. A different setting – a new but still artificial framework – facilitates the child's recovery and allows him or her to rebuild moral perceptions damaged by the abuse. A similar effect occurs when people with addiction problems move into group homes where they receive treatment and support. An even simpler example might be the external structures we commonly use to facilitate good manners and behavior: fences, the numbers we take while waiting at a bakery, rope-and-stanchion barriers that keep lines of people in order, and so on.

Of course, not all of the actions that build artifactual models are tacit; many are explicitly projected and planned. But imagine the people who first created these artifacts (the founders of group homes for addicted people, for example) – it is not unlikely that they created these entities simply and mainly "through doing," by creating new tacit templates of moral actions, rather than by following already established projects.

Moral Agents and Moral Patients

Because technological artifacts are designed, produced, and used in the human world, they are deeply interwoven with social interaction and, as a result, they have profound effects on what people do and how they do it. We can say, for example, that computers possess moral agency because

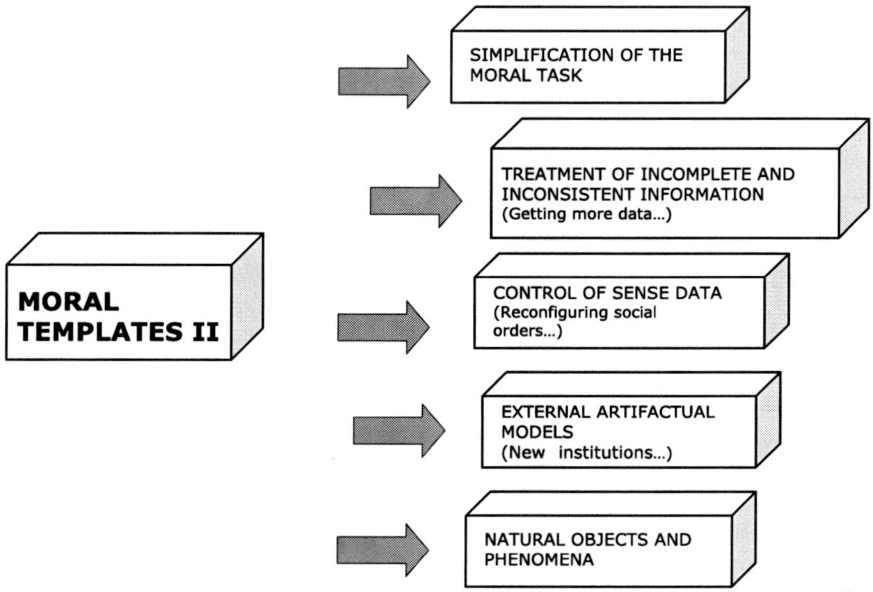

FIGURE 6.2. Conjectural moral templates II.

they (1) have a kind of intentionality and (2) can have moral effects on other entities (referred to later as "moral patients") – that is, they can benefit or harm beings that are capable of having their interests impeded or furthered. As Deborah Johnson puts it:

> Artifacts are intentional insofar as they are poised to behave in a certain way when given input of a particular kind. The artifact designer has a complex role here for while the designer's intentions are in the artifacts, the functionality of the artifact often goes well beyond what the designer anticipated or envisaged. Both inputs from users and outputs of the artifacts can be unanticipated, unforeseen, and harmful.[76]

One way to interrogate a moral problem is to think of the involved entities as subjects and objects of actions or, as some ethicists put it, as "moral agents" and "moral patients." Moral *agents* perform good or evil actions and are, therefore, sources of moral action, while moral *patients* are the objects of such actions. Floridi and Sanders posit that there can be several kinds of possible relationships between these two classes of agents:

1. They are disjoint (it is unrealistic to say that there is not at least one entity that qualifies both as an agent and as a patient);

[76] Johnson, 2004.

2. The second class can be a proper subset of the first;
3. The two classes intersect each other. Characterizations 2 and 3 are not very useful because they both require at least one moral agent that in principle could not qualify as a moral patient. (Only supernatural agents can fulfill this requirement: a God that affects the world but is not affected by the world, for example);
4. All entities that qualify as agents also qualify as patients and vice versa (standard position), or, finally,
5. All entities that qualify as patients also qualify as agents.[77]

This last statement may need a bit of revising, however: our discussion in the chapter 1 section "Ecology: 'Things' in Search of Values" suggests that animals are moral patients that cannot serve as moral agents. Generally speaking, nonliving "things" are considered to be similarly passive; like animals, both they and some artificial entities can be considered patients with intrinsic value – take the *Mona Lisa*, for instance. I contend, however, that unlike animals, some nonliving "things" – the internet, for example – are more than passive objects. Such things can be said to possess a sort of moral agency even though they lack the characteristics we usually associate with human agency: free will, full intentionality, responsibility, and emotion.

While this distinction between moral patients and moral agents may be correct and useful, it fails to recognize the dynamic aspects of moral delegation and externalization I have illustrated in this book. Indeed, moral delegation to external objects and artifacts does not take place simply because a given thing is supposed to possess a given set of properties intrinsically. For example, it is the way in which a thing dynamically interacts with humans and how they respond to it that renders an artifact like the *Mona Lisa* a moral patient. In this sense, my conception differs from the one that distinguishes moral patient from moral agent, which I consider to be too static.

This view fails to account for the process by which we continuously delegate and give (moral) value to the things around us. For example, the agent/patient distinction fails to explain why the first gift a teenager receives from a girlfriend or boyfriend might acquire great (intrinsic) value and become a simple moral patient, even if it is just a ragged old T-shirt. In such cases, value derives from neither the condition nor the cost of the present, but from the worth bestowed on it by the gift's recipient.

[77] Floridi and Sanders, 2004b. Carsten Stahl (2004) has recently investigated whether or not computers can be considered autonomous moral agents. Since computers cannot understand the information they store and manage, he says, they lack the basic capacity "to reflect morality in anything." He argues this point using an interesting and curious test called "the moral Turing test."

Moreover, there is an additional reason to prefer my conception of moral delegation. Consider, for example, the idea that animals should have rights on their own, a notion based on the claim that, like us, animals are capable of suffering. They are moral patients, and as patients they have to be respected. According to my view, we can further see this value attribution as a result of moral mediation. My view also shows that as we delegate new moral worth to animals, they can reveal to us previously unseen new moral features – of suffering, in this case, which then acquires a new value and a new extension. Animals play the role of moral mediators because they mediate new aspects of human beings' moral lives.

The agent/patient distinction, then, has some shortcomings: it is obvious that the moral agency of computers is not the same as that of human beings, and in this respect it is not different in kind from that of other technologies. Tom Powers has argued that while computers have a kind of external intentionality – one that, in the case of human beings, is expressed outside the human body through speech, written sentences, maps, and other designed artifacts – they cannot have *internal* intentionality in the way that people can. Instead, the agency of technological artifacts is closer to that wielded by human "surrogates" like tax accountants and estate executors.[78] This comparison elucidates the moral character of computer systems by showing that they have a kind of intentionality and have effects on moral patients, and hence that they are appropriate objects of moral appraisal. In these cases, we are faced with a kind of "mind-less morality."[79] The fact that some artifacts possess moral agency is yet another reason to seek new knowledge: we must construct suitable policies that allow us to "punish" problematic artifacts – that is, to modify, re-engineer, or remove them.

Moral Mediators

We have already established that the monumental moral challenges created by technology are making it necessary for us to think about the world in different ways, and I do not believe that concepts like the moral agent/moral patient paradigm are sufficient to give us the tools we need to achieve this goal. I contend that framing new moral problems by employing the idea of *moral mediators* is a much more fruitful way to proceed.

Much of the behavior we conduct through learned habits – the tacit templates of action described ealier – is devoted to building vast new sources of information and knowledge: external *moral mediators*. Let us return to our previous example of those human beings who create, in a

[78] Powers, 2004.
[79] Floridi and Sanders, 2004a.

tacit and embodied way, foster homes, which play a moral role: they facilitate a foster child's recovery as they allow him or her to rebuild moral perceptions that were damaged by previous abuse. Here, the artifact (the foster home) is an example of a moral mediator in the sense that it mediates – objectively, "over there," in an external structure – positive moral effects. Other moral mediators, however, are built by human collectives in a more conscious way, as in the case of some objectified rules and principles created with a particular goal in mind or already having established ways of producing moral effects (religious rites, for examples). Many complicated external moral mediators can also *redistribute* moral effort: they allow us to manipulate objects and information in a way that helps us to overcome the paucity of internal moral options (principles and prototypes, etc.) currently available to us. I also think that moral mediators can help to explain the "macroscopic and growing phenomenon of global moral actions and collective responsibilities resulting from the 'invisible hand' of systemic interactions among several agents at the local level."[80]

Using moral mediators is more than just a way to move the world toward desirable goals: it is an action that can play a moral role and therefore warrants moral consideration. We have said that when people do not have adequate information or lack the capacity to act morally upon the world, they can restructure their worlds in order to simplify and solve moral tasks. Moral mediators are also used to reveal latent constraints in the human/environment system, and these discoveries grant us precious new ethical information. Imagine, for instance, a wife whose work requires long hours away from her husband, and whose frequent absences cause conflict in their relationship. To improve their marriage, she restructures her life so that she can spend more quality time with her spouse, an action that can cause variables affected by "unexpected" and "positive" events in the relationship to covary with informative, sentimental, sexual, emotional, and, generally speaking, bodily variables. Before the couple adopted a reconfigured "social" order – that is, increased their time together – there was no discernible link between these hidden and overt variables; a new arrangement has the power to reveal important new "information," which, in our example, might come from a revitalized sex life, surprisingly similar emotional concerns, or a previously unrecognized intellectual like-mindedness.

A realigned social relationship is just one example of an external moral mediator; natural phenomena can also serve this purpose. In previous chapters, we considered the problem of "respecting people as things" as we discussed the potential of external "natural" objects to create new ethical knowledge. We have seen that endangered species serve as moral

[80] Ibid.

mediators, for example, when people define themselves as "endangered," for such comparisons provide us with new insights into other humans beings' self-perceptions.[81] In fact, many external things that have traditionally been considered morally inert can be transformed into moral mediators. For example, we can use animals, the earth, or cultural entities to identify previously unrecognized moral features of human beings, and we can employ external "tools" like writing, narrative, ritual, and, as we saw in the earlier examples of foster homes and group homes, institutions to reconfigure unsatisfactory social orders. Hence, not all moral tools are inside the head – many are shared and distributed in external objects and structures that function as objectified ethical devices.

While almost any sort of entity can help to mediate our moral outlook, certain technological artifacts can be considered *über* moral mediators – those equipped with artificial intelligence and the ability to be directly engaged in ethical reasoning and behavior. We must recognize the ethical ramifications not only of using such machines, but also of allowing them ethical autonomy vis-à-vis human users as well as other devices; in the process, we must assuage human fears about machine intelligence.[82] Developing ways to deal with artificial intelligence is a new field of research – called machine ethics – that involves many interesting topics: improving the interaction between artificial and natural intelligence systems by adding an ethical dimension to technological devices; using ethical strategies to enhance machine-to-machine communication and cooperation; developing systems that provide expert ethical guidance; establishing decision-making procedures for ethical theories with multiple prima facie duties that present conflicting perspectives; and assessing the impact of machine ethics on society.

External moral mediators of all kinds can function as components of a memory system that crosses the boundary between person and environment. For example, they transform the tasks involved in simple manipulations that promote further moral inferences at the level of model-based abduction. In the earlier case of the wife seeking moral protection of her marriage, she transforms it by manipulating her behavior so as to increase the quality time spent with her husband, and thus discovers new information that allows her to abduce new internal model-based ideas and/or her feelings – new motivating images, for example, or constructive emotions – about her husband and/or her marriage. When the everyday

[81] Cf. chapter 1.

[82] At the November 2005 AAAI symposium titled Machine Ethics (Anderson, Anderson, and Armen, 2005), it became clear that investigating these kinds of machines could reveal weaknesses in current ethical theories and further advance our thinking about ethics. For an informed taxonomy of most of the mediating ethical effects of machines, see the presentation given at the symposium by Moor (2005).

life of a previously abused child is manipulated by placing her with a foster family, for instance, the new setting is a moral mediator that can help her abduce new model-based internal experiences, images, emotions, or analogies through which she may be able to recalibrate her conceptions of adults, her past, and of abuse in general.

Actions executed through tacit templates can even enhance one's level of physical sensitivity: I can alter my bodily experience of pain by following the previously mentioned *control of sense data* template – that is, by unconsciously modifying the experience of my body and changing its relationships with humans or nonhumans in distress, I may, for instance, create new, empathic moral ways to help other beings. Mother Theresa's rich moral feeling and consideration of pain had certainly been shaped by her proximity to starving and miserable people and by her manipulation of both her own and their bodies. In many people, moral training is often related to these kinds of spontaneous (and sometimes "lucky") manipulations of their own bodies and sense data so that they build morality immediately and nonreflectively "through doing."

Throughout history, women have traditionally been thought to place more value on personal relationships than men do, and they are often regarded as more adept in situations requiring intimacy and caring. It would seem that women's basic moral orientation emphasizes taking care of both people and external things through personal, particular acts rather than relating to others through an abstract, general concern about humanity. The ethics of care does not consider the abstract "obligation" to be essential; moreover, it does not require that we impartially promote the interests of everyone alike. Rather, it focuses on small-scale relationships with people and external objects, so that, for example, it is important not to "think" of helping disadvantaged children all over the world (as men tend to aim at doing) but to "do" so when called to do so, everywhere.[83]

My philosophical and cognitive approach to moral model-based thinking and to morality "through doing" does not mean that this so-called female attitude, being more closely related to emotion, should be considered less deontological or less rational and therefore a lower form of moral expression. I contend that many of us can become more intuitive, loving parents and, in certain situations, learn to privilege the

[83] Moreover, both feminist skepticism in ethics and the so-called expressive collaborative model of morality look at moral life as "a continuing negotiation among people, a socially situated practice of mutually allotting, assuming, or deflecting responsibilities of important kinds, and understanding the implications of doing so" (Urban Walker, 1996, p. 276). Of course, this idea is contrasted with the so-called theoretical-juridical conception of morality. Johns (2005) includes a discussion of using sedation in the so-called critical care of patients who suffer severe emotional and physical distress after being attached to a piece of medical technology.

"taking care" of our children by educating our feelings – maybe by heeding "Kantian" rules.[84] The route from reason to feeling (and, of course, from feeling to reason) is continuous in ethics. Many people are suspicious of moral emotional evaluations because emotions are vulnerable to personal and contextual factors. Nevertheless, there are moral circumstances that require at least partially emotional evaluations, which become particularly useful when combined with intellectual (Kantian) aspects of morality.

Consequently, "taking care" is an important way to look at people and objects, and, as a form of morality accomplished "through doing," it achieves status as a fundamental kind of moral inference and knowledge. Respecting people as things is a natural extension of the ethics of care; a person who treats "nonhuman" household objects with solicitude, for example, is more likely to be seen as someone who will treat human beings in a similarly conscientious fashion. Consequently, using this cognitive concept, even a lowly kitchen utensil can be considered a moral mediator.

When I clean the dust from my computer, I am caring for the machine because of its economic worth and its value as a tool for humans. When, on the other hand, I use my computer as an epistemic or cognitive mediator for my research or in my didactic activities, I am considering its intellectual prosthetic worth. To make a case for respecting people as we respect computers, we can call attention to the values human beings have in common with these machines: (1) humans beings are "tools," albeit biological ones, with economic and instrumental value and, as such can be "used" to teach and inform others in much the same way we use hardware and software, so human beings are instrumentally precious sources of information about skills of various kinds; and (2) like computers, people are skillful problem solvers imbued with the moral and intrinsic worth of cognition.

COMPARING ALTERNATIVES AND GOVERNING INCONSISTENCIES

When dealing with any moral problem, we seek the moral considerations that are most closely related to that issue, but difficulties can arise when the considerations that seem most applicable derive from different and even inconsistent views, values, or principles. When considered together, the various values are sometimes irreconcilable and, as a result, lead to incompatible conclusions. What is the role of "coherence" in moral deliberation? Can we achieve the knowledge we need without it? This

[84] The role of the ethics of care in bioethics is illustrated in Carse, 1999.

section takes a methodological approach to comparing alternatives and governing inconsistencies.

This phenomenon of conflicting principles is evident in everyday human behavior: people often adopt different perspectives and frequently switch between them. Some of the values are outcome-oriented (utilitarian), some abstracted from the possible consequences of actions (obligations and duty), some are personal and agent-centered, and others are impersonal.[85] Many perspectives are mutually exclusive – the preference for capital punishment versus that for life in prison, for example – and we yearn for "objective" criteria that we can use to compare them.

Reasoning in terms of coherence is considered very important in justifying judgments of right and wrong, usually in contrast to a conception of right and wrong solely in terms of first principles.[86] To be perceived as right and legitimate, justice and moral judgment must be the product of reconcilable considerations applied during careful, coherence analysis. In the theory of justice, for example, reflective equilibrium is a state in which a thinker has reached a mutually coherent set of ethical principles, particular moral judgments, and background beliefs.[87]

The "ethical coherence" point of view, as illustrated by Paul Thagard, can delineate these problems.[88] We already know that when faced with a moral situation we have to consider (good) reasons and facts[89] and then make inferences using them. Let us consider the case of Paul Bernardo, a Canadian convicted of sexually torturing and killing two young women.[90] In accordance with Canadian law, Bernardo was sentenced to life in prison, but some people who generally consider capital punishment immoral held the inconsistent view that he deserved the death penalty given the especially horrific nature of the crimes. Such moral dissonance is an example of inconsistency permeating the belief system of just one individual. Let us now explore the problem at a societal level.

There are many reasons that can be given to support either capital punishment or life in prison:

1. *Deductive* arguments comprise propositions that express general "deontological" – Kantian – considerations and the consequent moral judgments, like the two opposing principles that capital punishment is wrong and that capital punishment is right in the case of particularly grisly crimes. The first principle entails the

85 Nagel, 1979.
86 DeMarco, 1994; Sayre-McCord, 1996.
87 Rawls, 1971.
88 Thagard, 2000.
89 Proved acts are always "reasons" in themselves, because they represent evidence we usually accept without hesitation.
90 The case is reported by Thagard, 2000, pp. 125ff.

statement that Bernardo's life should be spared, and the second that he should be executed. From the coherence point of view, certainly the first principle is coherent with the statement concerning a life sentence for the individual in question, Bernardo, and a positive constraint is established between the principle and the individual statement. (In the same sense, we can say that an axiom is coherent with the theorems derived from it, and vice versa: a deductive relationship expresses a "cognitive" coherence between propositions.) Hence, two different principles – capital punishment and life in prison – are in contradiction, or not coherent, and thus linked by a negative constraint.

2. *Explanatory* reasons, which evaluate "facts" and empirical hypotheses, are composed of propositions like hypotheses and evidence statements:

> Particular judgments such that Paul Bernardo should be punished depend on factual claims such as that he actually committed the crimes of which he is accused. General principles such as adoption of capital punishment can also be closely tied to factual claims: one common argument for capital punishment is that it is desirable as a deterrent to future crimes which depends on the empirical hypothesis that having capital punishment as a possible punishment reduces crimes of certain sorts. Evaluation of this hypothesis depends on a very complex evaluation of evidence. ... The principle that preventing crimes is good and the empirical hypothesis that capital punishment helps to prevent crimes together entail that capital punishment is good. These three propositions form a mutual constraining package.[91]

The three propositions are coherent, so that deductive coherence is connected to explanatory coherence: unlike the deontological principle, the empirical hypothesis requires support from empirical evidence, which, since we tend to accept the results of observations, enjoys a somewhat privileged status.

What happens when there exists an *alternative* legitimate moral perspective that is *inconsistent* with the previous one?

> An opponent of capital punishment might argue that killing innocent people is wrong, and that capital punishment sometimes kills innocent people, so that capital punishment is wrong. This entailment depends on the empirical hypothesis that innocent people are sometimes executed in countries and states that have capital punishment.[92]

[91] Ibid., p. 135.
[92] Ibid., p. 136.

For people who trust this last empirical hypothesis and the related evidence, the conclusion that capital punishment is wrong will be more coherent: consequently, the other (previous) "package" would not be legitimate because it seems (on the whole) less coherent than the present one. Let us remember that, analogously, "explanatory" coherence characterizes scientific reasoning, where one scientific theory is chosen over others because it furnishes the best (available) "explanation" of the evidence, and where the best explanation results from an overall coherence judgment.[93]

3. *Deliberative* arguments come from utilitarian considerations and add new factors not yet considered in the previous cases. If an action facilitates a consequence, then the action and the consequence are coherent, and a positive link is established. For example, killing Bernardo will facilitate the consequence that he will not murder again. Negative constraints are established when inconsistent perspectives are considered: we cannot execute Bernardo, for instance, and at the same time sentence him to life. Another argument could be that executing Bernardo will be cheaper than imprisoning him, and in that way taxpayers will bear less of a burden. The deterrence-based argument can be restructured in utilitarian terms: the prevention of murder will be the consequence of execution. We have already seen at the beginning of this chapter that in ethical decision making, "deliberative" considerations have to take into account all the people involved and not just one individual agent in order that the actions ultimately chosen are the ones that best serve the interests of all concerned. In our case, then, both the interests of Bernardo and those of the victims' families and anyone else involved must be considered. Of course, we must recall that deliberative coherence depends upon evaluating hypotheses about the nature of human beings and their society:

   > Deliberative coherence does not reduce to explanatory coherence, but depends on it in very useful ways that allow for the possibility of revising views about what to do. For example, the families of Paul Bernardo's victims may naturally want to see him killed, but whether execution would bring some relief from their grief is an empirical question. Without psychological evidence about the effects of executions in similar cases, we do not have grounds for saying whether an execution is really in the objective interests of the victim's families.[94]

4. *Analogical* reasons involve the use of analogies to modify or support moral conclusions – an example in the Bernardo case might be

[93] Thagard, 1992.
[94] Thagard, 2000, p. 131.

> comparing that incident to previous similar cases. If we compare capital punishment to killing a defenseless victim, an action that is considered to be wrong, then capital punishment can be valued as wrong. So the analogy institutes a coherence between the two cases, and a positive link is established: if we believe that killing is wrong, this belief is propagated through the positive link to the decision in favor of execution, which will also be considered wrong.

Paul Thagard and Karsten Verbeurgt propose a characterization of coherence as maximization of satisfaction of positive and negative constraints ("multiple constraint satisfaction"), so that a coherent problem can be stated by specifying a set of elements to be accepted or rejected along with sets of positive and negative constraints that incline pairs of elements to be accepted together or rejected together.[95] They have also provided a computationally tractable and psychologically plausible algorithm for determining the overall acceptance or rejection of elements in a way that maximizes coherence. According to the computational algorithm, the ethical assessment of capital punishment will depend on the overall combination of deliberative, explanatory, deductive, and analogical coherence.

It is more than clear that real human beings usually make ethical judgments by taking into account only one of these previously illustrated levels of moral reasons. For example, it is common to see utilitarians employing only what Thagard calls deliberative coherence, or Kantians disregarding consequences in favor of principles (deductive aspect). Human psychological and cognitive resources are limited, so it is mentally difficult to apply all the elements illustrated in Thagard's abstract model while simultaneously attempting to calculate and maximize the overall coherence of competing moral options. Put another way, real human beings can manage the previously illustrated coherence framework only in local areas; indeed, they are inclined to furnish immediate and simple conclusions that involve only one level of Thagard's abstract model.

I do not think that achieving coherence necessarily leads to the right moral decision. Of course, the "coherence" view just described is terrifically interesting because of its capacity to show the multidimensional character of ethical problems. Nevertheless, as I have illustrated in the previous chapter (the section "Owning Our Own Destinies"), I do not think that "coherence" in moral behavior is good in itself. Many people lack awareness of their own inconsistencies, and, moreover, coherence is knowledge-dependent: what appears coherent in a particular set of beliefs at one moment can, in light of new knowledge, later be revealed as inconsistent. Having accurate knowledge about our behavior, concerns,

[95] Thagard and Verbeurgt, 1998.

rights, and duties seems to be the most important task, for it exposes the inconsistencies of our perspectives and helps us to open new ones.

Balancing conflicting reasons is a fundamental challenge, however, and as Thagard points out, incoherent attitudes can, at times, help us to make far-seeing moral judgments. When we pay attention to the problems of another person, we can adopt a perspective that, while completely inconsistent with expectations we have for ourselves, is appropriate to that case. Cultural relativism teaches us that inconsistencies are sometimes good.[96]

I noted earlier that conflicts can crop up in any point of view, depending on how it is used to frame and interrogate an issue. Imagine, for example, a classroom in country A where a teacher is relating a famous story from ancient history that celebrates stamina, courage against the odds, and a willingness to die for one's country. The tale involves a fight between three Horatii of Rome and three Curiatii of Alba Longa, a battle decreed by Tullius Hostilius as a way to avoid a full-scale war between the two cities. In the story, all three Curiatii were wounded, and two Horatii were slain; the remaining warrior for Rome, Horatius, then managed to kill his weakened opponents. When he returned victorious to his city, his own sister, who had loved one of the Curiatii, wept rather than rejoiced, and he killed her for her disloyalty to Rome.

Then imagine that among the students hearing the story is an immigrant child from country B, who, because of his native country's culture, privileges allegiance to one's country over the prohibition against killing and, consequently, identifies positively with Horatius. This student will probably change his moral opinion, however, when the teacher moves from an abstract ideal to a concrete possibility by asking what would happen if the people of country A were to adopt the values demonstrated by Horatius. By considering how privileging loyalty to one's country over hospitality and friendship in the "real world" might affect the individual in question – here, the immigrant child – the teacher invokes outcomes in which the child's well-being is jeopardized and tolerance for him as a "foreigner" is eroded – such a scenario would, one assumes, engender a more tolerant attitude on the part of the student. This shift, of course, introduces an inconsistency in the boy's belief system, but if he abandons his previous radical commitment to his country and adopts the tolerant point of view, the inconsistency is eliminated.

Extending Rachels's conception of morality – guiding one's conduct by reason while taking into account the interests of others – helps us better to

[96] In chapter 5 (the section "Inconsistencies and Moral Behavior"), I used the example of my inconsistent judgments about myself and a colleague engaged in bureaucratic duties. I can personally hold two inconsistent moral attitudes, and they can both be correct because of the need to respect cultural diversity among human beings.

understand the methodological problems involved in producing and applying ethical knowledge, which, as I have said before, must be modernized to meet the challenges posed by a rapidly changing technological world. In this sense, knowledge is certainly a duty, not only when it comes to rational scientific knowledge, but also from the ethical point of view.

Let us use an analogy to abductive reasoning: one way to understand a given type of moral reasoning is to classify it as either creative or selective. If we generate brand new moral knowledge in order to deal with a situation, we are using creative moral reasoning, while if we choose an already-established principle or perspective from an encyclopedia of options, we are employing selective moral reasoning. Furthermore, this distinction holds for ethical inferences made using various approaches; it applies not only at the propositional and model-based level, for example, but also to "morality through doing." From this perspective, a more manageable description of moral knowledge and its application becomes possible: moral outcomes can be seen as related to the magnitude and strength of available ethical knowledge regarding certain needs, situations, and problems. Moreover, the awareness of the plurality of principles, codes, and attitudes does not lead to relativism.

Also important is the fact that the "morality through doing" framework sheds new light on the basic moral orientation of "taking care." This approach, which is traditionally associated with women, emphasizes engaging with human and nonhuman beings in a personal way rather than through some abstract concern for humanity or the world in general. My philosophical and cognitive approach to moral model-based thinking and to morality "though doing" reveals that this disposition's relationship to emotion does not render it less rational and deontological, nor does emotion's role mean that "taking care" is an inferior, lower level of thinking.

Another fruit of my methodological perspective is a new understanding of the role of coherence in ethical reasoning: my approach provides a very general model that conveys the multifaceted nature of comparing alternatives and governing inconsistencies. I believe the best moral deliberation results from creating or abductively selecting possible "reasons" (which, of course, may be propositional, model-based, hybrid, or a template of moral doing), then – at least ideally – choosing the most coherent set of reasons from the competing possibilities, and, finally, appropriately applying those reasons to the moral issue.

The next chapter will round out this chapter's discussion of several varieties of moral reasoning; chapter 7 takes on the very *nature* of ethical "reasons," which I believe can be clarified by taking an epistemological tour of concepts like abduction, manipulation, and "through doing."

7

Inferring Reasons

Practical Reasoning, Abduction, Moral Mediators

> The French have already discovered that the blackness of the skin is no reason why a human being should be abandoned without redress to the caprice of a tormentor. It may one day come to be recognized that the number of legs, the villosity of the skin, or the termination of the os sacrum, are reasons equally insufficient of abandoning a sensitive being to the same fate.
>
> Jeremy Bentham, *Principles of Morals and Legislation*

This last chapter – in part a twin of, and complementary to, the previous one – is intended to clarify some central methodological aspects of practical decision making and of ethical knowledge and moral deliberation. What is the role of reasons in practical decision making? What is the role of the dichotomy abstract versus concrete in ethical deliberation? What are ethical "reasons"? The first part of the chapter will reconsider and reevaluate some aspects of the tradition of "casuistry" and analyze the concept of abduction as a form of hypothetical reasoning that clarifies the process of "inferring reasons." As I have explained in the previous methodological chapter "Creating Ethics," I contend that morality is the effort to guide one's conduct by reasons, that is, to do what there are the best reasons for doing while giving equal weight to the interests of every individual who will be affected by one's conduct. I have also added (1) that it is necessary to possess good and sound reasons/principles applicable to the various moral problems, and (2) that we need appropriate ways of reasoning – inferences – that permit us to optimally choose and apply the available reasons.

The second part of the chapter will illustrate that "abduction" – or reasoning to explanatory hypotheses – is central to understanding some features of the problem of "inferring reasons." I contend that ethical deliberation, as a form of practical reasoning, displays many features of hypothetical explanatory reasoning (selection and creation of hypothesis, inference to the best explanation) as it is described by abductive reasoning in science. Of course, in the moral case we have reasons that support conclusions instead of explanations that account for data, as in epistemological settings.

To support this perspective, I propose a new analysis of the "logical structure of reasons," one that supports the thesis that we can look to scientific thinking and problem solving for models of practical reasoning. The distinction between "internal" and "external" reasons is fundamental: internal reasons are based on a desire or intention, whereas external reasons are, for instance, based on external obligations and duties that we may recognize as such. Some of these external reason can be grounded in epistemic mediators of various types. Finally, it is important to illustrate why it is difficult to "deductively" grasp practical reasoning, at least when we are aided only by classical logic; complications arise from the intrinsic multiplicity of possible reasons and from the fact that in practical reasoning we can often hold two or more inconsistent reasons at the same time.

The third part of the chapter, on "Abductive Reasoning and Epistemic Mediators," plays the role of a sort of appendix that describes abductive reasoning in scientific settings in the light of epistemology and cognitive science. The concepts of the non-monotonicity of reasoning, of model-based and manipulative abduction, and of epistemic mediators are illustrated. The reader can usefully exploit this part to deepen his understanding of the epistemological counterparts of those concepts that are used in the book for studying issues of ethical reasoning and knowledge.

GOOD MORAL HYPOTHESES/REASONS AND GOOD MORAL INFERENCES

In the previous chapter, I illustrated various kinds of ethical knowledge as applied to particular cases and situations. I contended, in keeping with James Rachels's "minimal" requirement, that agents must provide moral hypotheses based on *good reasons* and then apply those hypotheses to specific cases through the use of *good arguments*. This framework for ethical reasoning emphasizes three methodological issues:

1. From a cognitive perspective, there are many types of *moral hypotheses* that provide good reasons in moral deliberation and action: they can take the form of principles, rules, prototypes, previous analogical cases, examples, images, visualizations, "educated" feelings, metaphors, narratives, and so on; we have seen that even "implicit" hypotheses can inform a kind of "morality through doing." I also stressed the central role of what I have called "moral mediators,"

which result in moral deliberation as a kind of action that is mainly manipulative rather than derived from abstract thinking. As the moral agent manipulates objects and situations in both human and natural environments, he or she achieves morality immediately "through doing" rather than "through feeling" or "through reasoning."

2. Ethical *knowledge* must be improved or, in some cases, created anew, so that we may deal effectively with important and ambiguous new problems. Those challenges include not only the many ecological, biotechnological, and cybernetic challenges I outlined in chapters 1, 2, and 4, but also the phenomenon of hybrid people – those strange new human creatures that, as we saw in chapter 3, have been generated by the recent revolution in information and bioengineering technologies. All these cases feature rapidly changing environments and situations, which must be recast and reenvisioned if we are to understand and manage them well.
3. Without an effort of reasoning (and of adequate moral feeling), even well-intentioned moral actions by individuals or collectivities can be insufficient, incorrect, and/or futile. Hence, ethical creativity involves more than just rich *moral knowledge* (good reasons) combined with a strong will to do right things: it also requires good *moral inferences* that are verbal-propositional as well as model-based and "through doing" (possibly abductive, nonmonotonic, defeasible).

The Abuse of Casuistry, Albert Jonsen's and Stephen Toulmin's interesting study of the history of moral reasoning, underscores the positive aspects of casuistry, an ethical approach that has been maligned for centuries.[1] When faced with an ethical problem, practitioners of casuistry do not make choices based on an overarching moral code that they consistently apply to such situations; rather, they find guidance by sifting through previous moral cases and then engaging in complex verbal argumentation during which they examine the present issue and compare it to the earlier cases. Generally speaking, engaging in casuistry is related to the act of diagnosis, and just as making diagnoses requires physicians to have a wealth of medical knowledge, so does casuistry depend on a familiarity with moral issues. The casuistic agent benefits from having wide knowledge about a variety of ethical cases – greater knowledge yields more possible parallels to the case at hand – and consequently, the approach has been most commonly associated over the centuries with "professional thinkers" such as priests, monks, theologians, and

[1] Jonsen and Toulmin, 1988.

philosophers – the so-called casuists.[2] This type of reasoning was widespread among the men who held those positions during the Middle Ages, when casuistry was strongly linked to Catholic scholars, the Jesuits in particular.

Casuistic arguments were often so complex and impenetrable, however, that they were sometimes viewed as a method of subterfuge, a sort of confuse-and-conquer approach to moral persuasion; indeed, many such arguments have been so complicated and obscure that they seem to have produced anything but equilibrated moral judgment. Three hundred years ago, Blaise Pascal cast disrepute on casuistry in his famous *Provincial Letters* (1657) when he contended that the practice was in fact rife with corruption.[3] The objects of his attack were the Jesuits, whom he believed had created a rather too elastic idea of Christianity in their attempts to attract new converts, win back Protestants, and procure the favor of the wealthy and powerful. In Pascal's view, the Jesuits' prolix, jargon-laden writings about ethical cases allowed the order to circumvent certain inconvenient abstract moral rules and principles.

The associated doctrine of probabilism expresses the ethical dignity of the so-called probable opinions in casuistry. In this light, of course, every opinion could qualify as probable, generating an unmanageable surfeit of possibilities, as Pascal and almost all commentators on casuistry over the last three centuries have lamented.

AN OLD-FASHIONED DICHOTOMY: ABSTRACT VERSUS CONCRETE

This section is devoted to the tradition of casuistry. Indeed, it is useful to remember that moral judgments, and judgments in general, are not only "directly" and rigidly derived from well-established rules and principles, but may also be derived through the complicated verbal argumentation of "practical reasoning," in part based on what cognitive scientists today would call "case-based reasoning," and in part characterized by a careful attention to circumstances, concrete aspects, and possible exceptions. I think some aspects of casuists' methodology can be vindicated: for example, the emphasis on circumstances and exceptions; the importance of "probable" opinion and of the multiplicity of ethical "reasons"; the role of analogical reasoning; and the need to improve, to modify (and to create new) ethical knowledge when faced with puzzling cases. The

[2] This kind of reasoning is very similar to a legal argument based on precedents, where a judge who wants to make a ruling that departs from precedent must justify this opinion by showing not only why the case at hand is similar to previous cases, but also, crucially, why it differs from them.

[3] Pascal, 1967.

rapidly changing traits of current technologies compel us to continually face new or unexpected situations and problems, so the practical reason we adopt to deal with them must be particularly flexible.

Blaise Pascal's attack was so effective that it was a primary factor in the decline and disappearance of casuistry. The ethical method may, however, offer the modern reader some useful insights, as Jonsen and Toulmin suggest in their book, where they note the flexibility of casuistry and see it as much better suited to the exigencies of practical reasoning than is rigid "deductive" moral judgment, which is "geometrical, idealized, and necessary," a dogmatic descendent of rules. Casuistry, they write, can be thought of as a kind of "informed prudence," an heir to Aristotle's "concrete, practical, and presumptive" *phronesis*, since the casuist focuses on *circumstances*, on exceptions, and on substantive aspects of real-life moral cases.

Jonsen and Toulmin note similarities between casuistry and physicians' "diagnostic reasoning," wherein nondeductive procedures and concrete cases are more important than abstract medical laws, rules, and classical deductive reasoning. While Pascal decried this kind of reasoning as slippery, the authors see it as disciplined and specific, and they contrast it with thought processes not typically used in ethics: forms of "analytic" scientific reasoning. These approaches may be appropriate for dogmatic ethical reasoning – when abstract rules and principles are rigidly applied to, say, relatively simple problems – but they are often useless in the face of extreme and particularly difficult ethical cases.

Comparing ethical deliberation and diagnostic reasoning is useful and appropriate; we currently know that diagnostic reasoning can be analyzed as one example of a nonclassical form of reasoning called abduction.[4] Abductive reasoning is a frequently used form of epistemological reasoning in the creation and selection of hypotheses as well as in other theoretical and practical settings; in addition, the fields of logic, cognitive science, and artificial intelligence have produced several new rational models that are especially useful in identifying features of diagnostic reasoning in medicine.[5]

I believe, however, that using diagnosis metaphorically to underscore the specificity of casuistic reasoning – as Jonsen and Toulmin do – is epistemologically old fashioned and, in any case, does not reveal diagnostic reasoning to be more flexible than analytic scientific rationality. Abduction, a widely used form of human reasoning, is "not deductive," of course, at least in terms of classical logic (see the discussion later in this chapter), but this does not mean that it cannot be modeled in deductive

[4] I have described abductive reasoning in *Abduction, Reason, and Science* (Magnani, 2001a), and I will summarize its main characteristics later in this chapter.

[5] Diagnosis is an expert "selection" of suitable hypotheses that seek to satisfactorily explain data.

way. Many rational (and logical) models of medical diagnostic abductive reasoning have been constructed, as have several computational models that I call "automatic physicians."[6] Clinical medicine is certainly a practical enterprise, but it is not necessarily an example of reasoning that is "specie-specific" to a good moral agent.[7] I think that in morality, classical inferences – abstract and "deductive" – can give rise to good or bad ethical judgments in the same way that flexible abductive ones can. Moreover, practical reasoning is not unique to diagnosis or ethical deliberation – it is widely used in some manipulative scientific procedures that require, for example, creativity and model-based reasoning, as I will illustrate later in this chapter.

It seems to me that the analytic method in science is not the antithesis of the diagnostic method in casuistry, and, consequently, the dichotomy established by Jonsen and Toulmin must be rejected. After all, a diagnosis is not in itself a guarantee of a more appropriate ethical reasoning. Like any poorly framed or articulated argument, verbose, undisciplined casuistic reasoning can give rise to errors like those Pascal pointed out hundreds of years ago.[8] At present, however, there is greater epistemological, cognitive, and scientific awareness of defeasible and non-monotonic inferences, and, consequently, we now have more rigorous ways to test practical reasoning than relying only on the merely argumentative framework of classical casuistry. For example, the connectionist model of ethical deliberation in terms of "coherence" discussed at the end of the previous chapter is an example of this kind of "methodological" awareness.[9] Moreover, such a model supplies flexible, rational frameworks that can accommodate various deliberations and inferences, from those that are abstract (deductive), to those that are analogical and explanatory, to still others that are a blend of approaches. It facilitates moral reasoning about concrete and complicated cases not only for "professional thinkers," as is generally the case with casuistry, but for the nonspecialist as well.

Positioning analytic and diagnostic methods as opposites, as Jonsen and Toulmin do, creates an epistemological state that can easily give way to irrationalism. Celebrating practical inferences only seems to promise an orderly way to approach moral deliberation that would distinguish it a priori from the methods typical of science. Also, it seems Jonsen and Toulmin suggest that applying scientific methods to problematic ethical cases is ineffective, but this view employs a rather narrow idea of scientific approaches – it neglects central aspects of scientific reasoning like

[6] Magnani, 2001c.
[7] Jonsen and Toulmin, 1988, pp. 36–42.
[8] Ibid., pp. 42–46.
[9] See chapter 6, the section "Comparing Alternatives and Governing Inconsistencies."

discovery and theory comparison, which involve inferences that are structurally "similar" to those of both ethical deliberation and diagnostic reasoning.

For example, Jonsen and Toulmin correctly focus on the role of analogical or, to use their word, "paradigmatic" reasoning and point out that case-based reasoning is commonly used in diagnosis and casuistry. They do not, however, mention the fact that analogical and case-based reasoning are just two of many kinds of model-based reasoning employed in ethics as well as in various types of scientific reasoning and scientific creative processes. Moreover, we should also consider the inferential nature of other important types of thinking, such as those discussed in the previous chapter: visualizations, imagination, emotions, morality through doing, and moral templates. But it is important to remember that *The Abuse of Casuistry* was published in 1988, and it is only through recent advances in logic, artificial intelligence, and cognitive science that many kinds of thinking associated with practical reasoning have been revealed to be equally relevant to other fields, like science. In the present book, I hope I have been able to bridge this gap by explaining some logico-epistemological and cognitive issues I believe to be very important when dealing with practical and ethical reasoning.

THE MODERNITY OF CASUISTRY

Jonsen's and Toulmin's book usefully illustrates classic examples of performances by effective casuists: the cases of usury (which involve the difficult task of judging the morality of many ambiguous cases of economic activity), equivocation, and insult. All the cases highlight the necessity of updating moral knowledge so that it may keep pace with shifts in social and economic life; they also clearly teach us the importance of expanding our ethical knowledge and enhancing its inferential and argumentative power in the face of evolving technologies like those I have described in this book. As the examples trace the steps of the successful casuists' reasoning, it becomes clear that the casuists skillfully managed inconsistencies, reconciled conflicting applications of different moral hypotheses, deftly juggled concrete data and circumstances, and adroitly handled many other challenges posed by the technological circumstances described on many pages of this book. While casuistry may have some drawbacks, I still think it can suggest some methodological devices to us, methods that we can adopt in real-life situations when we ourselves are thwarted by a lack of moral knowledge or when we must sort through available information that seems to be riddled with conflicts.

Modifying Reasons, Principles, and Rules

A first example is related to the analysis of usury given by classical casuists since the Middle Ages. Jonsen and Toulmin contend that Islamic fundamentalists have resurrected medieval objections to the charging of financial interest as part of a more extended attack on Western influence and that they seek different ways of financing commerce and industry that, in their eyes, do less violence to Islamic society. They consider international loans from Western governments and banks to be exploitive but wish to find and retain elements of capitalism within their domestic economies as a way of promoting development within the family: "So the medieval debate about the clever new forms of contract, aimed at circumventing the moral objection to interest, is being repeated in contemporary Islam, in the hope of squaring the needs of commerce with the traditional injunctions of the *Sharī ya*."[10]

Here we see that using an old financial practice in a new context (modern Islam) generates problems; difficulties arise when loans are made between countries with different cultures. The medieval conflict between "moral" investing and immoral money lending acquires new relevance. Jonsen and Toulmin contend that if casuistic reasoning can be applied to the debate about usury, then a modern moral reasoning related to similar puzzling cases can be derived from that case.

Simply applying a general principle against usury is not particularly productive, for it limits opportunities for commerce between Muslim nations and the rest of the world; instead, new ways of conducting business must be considered. The underlying lesson here is that the concrete case – the seemingly irreconcilable conflict between cultures – takes agents beyond the reach of rules and compels them to take into account a particular set of circumstances – the fact that there are other commercial practices that are acceptable in Islamic business communities. In chapter 6, we considered a similar situation when we explored the "sanctity of human life" idea: it is not always a "good" principle to use when deciding how (or whether) to treat extremely premature infants with modern medical techniques, techniques that can often be very useful but can also have unacceptable side effects for both the children and their families. In both the usury problem and the infant treatment case, abstract rules must be suitably modified to fit particular circumstances. In the field of politics the moral presumptions also remain as open to refusal – for example, in the puzzling case of minorities who justify lying and murdering by appeal to the right of self-determination.

[10] Jonsen and Toulmin, 1988, p. 310.

Double Effect

Casuists warn against the problem of the double effect – when, for example, a pregnant woman sustains a serious injury and doctors are able to save either the mother or the fetus, but not both, and must choose which course of treatment to pursue. We encountered this problem in chapter 2 when dealing with biotechnological innovations: in the imminent technological future, we will have to decide whether to allow the use of cloning to enable some couples to have children, or to prohibit the practice in keeping with the Kantian maxim of not treating people as means.

Respecting People as Things

When we view killing as morally justifiable if it is carried out in defense of honor or property (as has recently occurred – lawfully – in some Western countries), we are faced with another puzzling phenomenon casuists have already considered. Indeed, we know that human beings sometimes attribute too much moral and "intrinsic value" to honor and property: does honor, an abstract concept (a "thing") count more than the biological life of a human being? Does property, also just a "thing," count more than a human life, so that when our land, house, or other belongings are threatened, we may ignore the prohibition of killing? It is clear that this last problem relates to the moral and theoretical issues I have illustrated in this book, and it brings us back to the concept of "respecting people as things." I hope I have clearly conveyed my belief that, paradoxically, acknowledging our respect for things here, for property and the abstract idea of honor, can teach us how to ascribe new value to human beings. In the case of honor and property, we have to reclaim the excess of intrinsic value we have externalized in order to recover the respect for human life.

Creating New Moral Knowledge

Technology creates totally new situations, from cloning to pollution to threats against privacy, freedom, and human dignity, as I have illustrated in the previous chapters. In many instances, we lack the appropriate moral knowledge necessary to deal with such unprecedented challenges, because we have not yet established a body of successfully solved cases from which we can draw analogues for ethical reasoning. In the last chapter of their book on casuistry, Jonsen and Toulmin present the highly unusual case of a married couple – with children – in which the husband, with the help of modern surgical technology and medical treatment, changes gender. The changes he underwent

> ...undercut some of the most basic *factual* assumptions about the married state. Specifically, both the idea and the institution of "marriage" take it for granted that any marriage initially contracted between a man and a woman will remain, for so long as it lasts, a marriage between a man and a woman. When we face the possibility of a marriage in which, say, the husband turns into a woman, issues arise about which neither the legal nor the moral precedents can give us clear or unambiguous guidance.[11]

The main issue here is that the everyday assumptions built into the term *marriage* are deeply shaken, and there are no tried-and-true methods for managing such situations. While a few human communities have adopted strategies that might help to clarify dramatic cases like this one, developing the widespread ability to deal with such issues will require continuous moral discussions. Engaging in such discussions will prepare us for new challenges and help us to avoid the disorientation and conflict that typically accompany an unprecedented situation, as was the case with the casuist solution to the problem of usury.

The casuists' emphasis on concrete cases and on continually modifying our moral knowledge (and on "creating" new moral frameworks) is undoubtedly modern: I have stressed throughout this book that recent technologies have made (and will continue to make) novel situations and problems unavoidable. I hope this book, by suggesting the new moral approach of "knowledge as duty" and of "respecting people as things," contributes to the much-needed discussion of the problems generated by technological challenges.

Does casuistry still need to be proven valuable? I would say only partially, because there are still things we don't understand. The following list summarizes several reasons not to discount casuistry:

1. We need not avoid casuistry because of a rhetorical prolixity that could lead to fallacies, mistakes, and pernicious relativism. We have already illustrated that practical inferences follow patterns of reasoning that have recently been studied and clarified from the rational point of view; consequently, these templates can be considered reliable and are therefore unlikely to lead to mistakes and fallacies.
2. We must not assume that focusing on concrete cases rather than on abstract theories, as we do in diagnosis and casuistry, involves a "special" kind of reasoning that works for moral reasoning but is inappropriate for rational modern science. I strongly believe that when we rationally engage in moral reasoning, it can lead us to inferences that result in the "good reasons and good argumentation"

[11] Ibid., pp. 318–319.

described earlier. I am not inclined to agree with the following dichotomy: "the outcome of our inquiry confirms what Aristotle taught long ago: that ethical arguments have less in common with formal analytic arguments than they do with topical or rhetorical ones. They are not concerned with theoretical relations *internal to* a system of concepts but with relating and applying those concepts *outwardly*, to the world of concrete objects and actual states of affairs."[12] Topical and rhetorical arguments are fallacy-prone and can lead to erroneous conclusions, but the rigid abstractions of "formal analytic arguments" can lead to unjust conclusions, even if they are sound from a logical point of view. Both can be bad in ethical reasoning. Moreover, ethical knowledge is not more particular and individual-oriented than scientific knowledge; both concrete and abstract aspects play important roles in each field.

3. Unlike its old-fashioned version, modern casuistry makes use of images, visualizations, and "educated" feelings, and it values the ability of implicit templates to substantiate "morality through doing"; in such cases, people reap the benefits of what I have called "moral mediators." For example, Jonsen and Toulmin refer to personal preferences in casuistry as merely "subjective" and "idiosyncratic," but at the same time they recall that casuists always stressed the importance of the "informed" conscience.[13] I contend that only "trained" moral feelings that acquire rational connotations can play a legitimate role in ethical deliberation and action, and this idea is more in tune with the casuists' emphasis on the "informed" conscience. In this sense, "informed conscience might be intensely personal, but its primary concern was to place the individual agent's decision into a larger context at the level of actual choice: namely the moral dialogue and debate of a community. ... 'liberty of conscience' never meant the right to take up a personal moral position that ran in the face of the general agreement of reflective scholars and doctors."[14]
4. It is certainly possible that an amended casuistry could suggest not only methodological but also pragmatic solutions. It could represent a valuable model for managing moral issues and problems in some modern ethical collectivities (those that are more or less official and/or have the authority of state institutions) that group together various kinds of professionals – ethicists, philosophers, scientists, general workers, politicians, and so on.

[12] Ibid., p. 327.
[13] Ibid.
[14] Ibid., p. 335.

Because it emphasizes argumentation and negotiation between conflicting parties, it could also create a useful paradigm for the way the collective itself is organized.

THE LOGICAL STRUCTURE OF REASONS

This section, which is in a sense the methodological core of the book, illustrates that "abduction" – that is, reasoning to hypotheses – is central to the problem of "inferring reasons" in ethical reasoning and deliberation. In abduction, we usually base our guessing of hypotheses on incomplete information, so we are then faced with nonmonotonic inferences: we draw defeasible conclusions from incomplete information, and these conclusions are always withdrawable. It is in this sense that abductive reasoning can constitute a useful model of practical reasoning: ethical deliberations are always adopted on the basis of incomplete information and on the basis of the selection of particular abduced hypotheses that play the role of "reasons." Hence, ethical deliberation shares some characteristics with hypothetical explanatory reasoning as it is typically illustrated by abductive reasoning in scientific settings. To support this perspective on the "logical structure of reasons," I will provide an analysis based on the distinction between "internal" and "external" reasons and on the difficulties involved in "deductively" grasping practical reasoning, at least with the only help of classical logic alone.

When I introduced the methodological problems of ethical deliberation in the previous chapter, I contended, following Rachels, that morality is the effort to guide one's conduct by reasons, that is, to do what there are the best reasons for doing while giving equal weight to the interests of each individual who will be affected by one's conduct. I added that (1) we need good and sound reasons/principles to apply to the various moral problems, which will call our attention to and encourage evaluation of (good) arguments for opposing moral views, and (2) we need appropriate ways of reasoning – inferences – that allow optimal application of the available reasons.

I also said that in *abductive reasoning* (see the section "Abductive Reasoning and Epistemic Mediators" later in this chapter), which is used to form explanatory hypotheses, we usually base our hypotheses on incomplete information, and that we are then faced with *nonmonotonic* inferences: we draw defeasible conclusions from incomplete information.[15]

[15] It must be noted that explanatory reasoning can be causal, but explanations are also based on other aspects. Thagard (1992, chapter 5) illustrates various types of explanations (deductive-nomological – in the so-called neopositivist tradition of philosophy of science – statistical, causal, analogical, schematic).

From this perspective, abductive reasoning also constitutes a model of practical reasoning: we adopt ethical deliberations based on incomplete information and on particular abduced hypotheses – guesses – that serve as "reasons." Hence, I contended that as a form of practical reasoning, ethical deliberation shares some characteristics with the hypothetical explanatory reasoning (the selection and creation of a hypothesis, inference to the best explanation) that occurs during abductive reasoning in scientific and diagnostic settings.[16] Of course, in moral cases we have reasons that support conclusions instead of explanations that account for data, as in epistemological settings.

"The Logical Structure of Reasons" is the title of chapter 4 in John Searle's book *Rationality in Action*.[17] I plan to use Searle's conceptual framework to give us a better understanding of the precise nature of "reasons" in ethics. While Searle deals with rational decision making, many of his conclusions appear to be appropriate for ethical cases as well.

By criticizing the classical model of rational decision making (which always requires the presence of a desire as the condition for triggering a decision), Searle establishes the fundamental distinction between *internal* and *external* reasons for action: internal reasons might be based on a desire or an intention, for instance, while external reasons might be grounded in external obligations and duties. When I pay my bill at a restaurant, I am not doing so to satisfy an internal desire, so this action does not arise from internal motivations; instead, it is the result of my recognition of an external obligation to pay the restaurant for the meal it has provided. Analogously, if an agent cites a reason for a past action, it must have been the reason that the agent "acted on." Finally, reasons can be for future action, and this is particularly true in ethics, where they do not always trigger an action – in this case, however, they must still be able to motivate an action: they are reasons an agent can "act on."

Searle's anticlassical "external reasons" will probably not seem strange to readers of the present book: I have often stressed the fact that human beings delegate cognitive roles (and moral worth) to external objects that consequently acquire the status of deontic moral structures. This also occurs when we articulate ideas in verbal statements – promises, commitments, duties, and obligations, for example – that then exist "over there," in the external world. Imagine the deontic role that concrete buildings (for instance, buildings whose shapes restrict the routes that people can follow) or abstract institutions (for example, constitutions that compel us to respect the equality of citizens) can play in depicting duties and commitments we can (or have to) respect. Human beings are

[16] Cf. Magnani, 2001a.

[17] Searle, 2001.

bound to behave in certain ways as spouses, taxpayers, teachers, workers, drivers, and so on. All these external factors can become – Searle says – reasons/motivators for prior intentions and intentions-in-action of human beings.

In chapter 1, I illustrated that many things around us are human-made, artificial – not only concrete objects like a hammer or a PC, but also human organizations, institutions, and societies. Economic life, laws, corporations, states, and school structures, for example, can also fall into that category. We have also projected many intrinsic values on things like flags, justice rituals, and ecological systems, and as a result, these external objects have acquired a kind of autonomous automatism "over there" that conditions us and distributes roles, duties, moral engagements – that is, that supplies potential "external reasons." Nonhuman things (as well as, so to say, "non-things" like future human beings and animals) become moral clients as well as human beings, so that current ethics must pay attention not only to relationships between human beings, but also to those between human and nonhuman entities.

Moreover, in the "Moral Mediators" section of the previous chapter, we saw how external things that are usually considered morally inert can be transformed into moral mediators that express the idea of a distributed morality. For example, we can use animals to highlight new, previously unseen moral features of other living objects, as we can do with the Earth or with (non-natural) cultural objects; we can also use external "tools" like writing, narratives, others persons' information, rituals, and various kinds of institutions to morally reconfigure social orders. Hence, not all moral tools are inside the head like the emotions we experience or the abstract principles we refer to; many are "over there," even if they have not yet been identified and represented internally, distributed in external objects and structures that function as ethical devices available for acknowledgment by every human agent. These delegations to external structures – which are thus transformed into moral mediators – encourage or direct ethical commitments, and, as I explained many times in chapter 3, they favor the predictability in human behavior that is the foundation for conscious will, free will, freedom, and the ownership of our own destinies. If we cannot anticipate other human beings' intentions and values, we cannot ascertain which actions will lead us to our goals, and authoring our own lives becomes impossible.

Let us return to the role played by reasons in ethical reasoning. Intentional states with a propositional content have typical *conditions of satisfaction* and *directions of fit*.

1. First, mental and linguistic entities have directions of fit: for example, a belief has a *mind-to-world* direction of fit. For example, if I believe it is raining, my belief is satisfied if and only if it is

raining "because it is the responsibility of the belief to match an independently existing reality, and it will succeed or fail depending on whether or not the content of the belief in the mind actually does fit the reality of the world."[18] On the other hand, a desire (or an order, promise, or intention) has a *world-to-mind* direction of fit: "if my belief is false, I can fix it up by changing the belief, but I do not in that way make things right if my desire is not satisfied by changing the desire. To fix things up, the world has to change to match the content of the desire."[19]

2. Second, other objects (not mental and not linguistic) also have directions of fit similar to those of beliefs. A map, for example, which may be accurate or not, has a *map-to-world* direction of fit, whereas the blueprints for a house have a *world-to-blueprint* direction of fit because they can be followed or not followed.[20] Needs, obligations, requirements, and duties are not in a strict sense linguistic entities, but they have propositional contents and directions of fit similar to those of desires, intentions, orders, commitments, and promises that have a *world-to-mind, world-to-language* direction of fit. Indeed, an obligation is satisfied if and only if the world changes to match the content of the obligation: if I owe money to a friend, the obligation will be discharged only when the world changes in the sense that I have repaid the money.

When we applied the moral principle of the *wrongness of discriminating against the handicapped* to the case of Baby Jane Doe, we resorted to a kind of "external" reason that we had to "internalize" – that is, recognize as a reason worth considering as we sought to orient our moral actions concerning the girl's life.[21] If we had instead used strong personal feelings like pity and compassion to guide our reasoning, we would have decided for or against the operation based on a completely "internal" reason. We have to note, of course, that external reasons are always observer-relative.[22] It is only human intentionality that furnishes meaning to a particular configuration of things in the external moral or nonmoral world. The objective fact that, say, I have an elevated white blood cell count acquires a direction of fit that is a direction for action only when related to a human being's interpretation (for example, only "in the light" of a diagnosed disease can this fact trigger a decision for therapy).

[18] Ibid., p. 37.
[19] Ibid., p. 38.
[20] Ibid., p. 39.
[21] Cf. the previous chapter, the section "Creating and Selecting Reasons."
[22] Cf. the previous chapter, the section "Moral Emotions."

Searle also discusses the so-called collective intentionality that enables people to create common institutions such as those involving money, property, marriage, government, and language itself, an intentionality that gives rise to new sets of "conditions of satisfaction," duties, and commitments. From my perspective, I say these external structures have acquired a kind of delegated intentionality because they have become *moral mediators*; they have acquired a kind of moral "direction," as I have indicated in the previous chapter.[23] I have already illustrated that in those cases, when we have to deal with a moral problem through moral mediators, evaluating reasons of any kind immediately involves manipulating nonhuman externalities in natural or artificial environments by applying old and new behavior *templates* that exhibit some uniformities. This moral process is still hypothetical (abductive): these templates are embodied hypotheses of moral behavior (either pre-stored or newly created in the mind-body system) that, when appropriately employed, make possible what I have called a moral "doing."

We must remember that I contend that external moral mediators are a powerful source of information and knowledge; they redistribute moral effort by managing objects and information in new ways, and as a result, they transcend the limits created by the poverty of moral options immediately represented or found internally – those options discovered, for example, by merely applying internal/mental moral principles, utilitarian envisaging, and model-based moral reasoning like emotions.

It follows from the previous discussion that many entities can play the role of deontic moral structures. As anticipated in the first section of chapter 4, this fact can lead to a reexamination of the concept of duty. From this perspective, duties can also be grounded in trained emotional habits, visual imagery, embodied ways of manipulating the world, exploitation of moral mediators – as we have just seen, endowed with a sufficient ethical worth in a collective.

The Ontology of Reasons

What are these "reasons" that, following Searle, are the basis of rational actions and, in the Baby Jane Doe case, the basis of moral action? A reason answers the question "why?" with a "because"; it can be a statement, like a moral principle, as in the answer to "Why should we perform surgery on Baby Jane Doe?": "Because of the wrongness of discriminating against the handicapped." In reality, reasons are "expressed" by the statements – *explanations* – insofar as they are *facts* in the world (the fact that it is raining is the reason I am carrying an umbrella). They are also

[23] Cf. the previous chapter, the section "Being Moral through Doing: Taking Care."

represented by *propositional intentional states* such as desires (my desire to stay dry is the reason I am carrying the umbrella), and, finally, by *propositionally structured entities* such as obligations, commitments, needs, and requirements, as in the case of our moral "principle" "the wrongness of discriminating against the handicapped." All good reasons explain, and all explanations give reasons. Searle also distinguishes between reasons that justify my action and thus explain why it was the right action to perform, and reasons that explain why in fact I performed it.

1. First of all, in rational decision making, when we must provide a reason for an intentional state, we have to make an intelligent selection from a range of reasons that exist either internally or externally – in the latter case, we must take the external reason, recognize it as good, and internalize it. With respect to my ideal of an ethical deliberation sustained by "reasons," I can affirm that it is not unusual for the "deliberator" to have limited knowledge and inferential expertise at his or her disposal. For instance, she may simply not have important pieces of information about the moral problem she has to manage, or she may possess only a rudimentary ability to compare reasons and ascertain data. I have often said in this book that ethical reasoning is defeasible: because it is impossible to obtain all information about any given ethical situation, every instance of moral reasoning occurs without benefit of full knowledge, so we must remember that any reason can be rendered irrelevant or inappropriate by new information. Generally speaking, as illustrated earlier, these reasons can take three different forms: external *facts* in the world, such as empirical data; internal *intentional states* such as beliefs, desires, and emotions; and *entities* in the external world like duties, obligations, and commitments with an upward direction of fit (world-to-mind). External facts must be internalized and "believed," while external entities must be internalized and adopted ("recognized") as good and worthy of consideration. The same happens in the case of rational moral deliberation.
2. Second, as I observed in the previous chapter, we must remember that maintaining a flexible, open mind is particularly important when we lack the ethical knowledge necessary to confront new or extreme situations.

It is this second idea that returns us to the idea of "knowledge as duty" from chapter 4. In our technological world, it is our duty to produce and apply updated ethical knowledge just as it is to gather and implement other kinds of knowledge – "scientific" knowledge, for example. In the ethical case, we stress the importance of selecting good reasons in terms of available principles, facts, and information by improving the methodological awareness of the main cognitive processes involved. I maintain that appropriate ethical knowledge and proper moral reasoning are the basic conditions for maintaining freedom and taking responsibility for our actions, which I have indicated in the previous chapter are the main traits of a moral life.

When evaluating an ethical case, we have at hand all the elements of rational moral decision making: the problem we face, the "reasons," and the agents involved. Every reason, Searle says, contributes to a "total reason" that is ultimately a composite of every good reason that has been considered – beliefs, desires, obligations, and facts, for example. First, as already observed, rationality requires the agent to recognize the facts at hand (I have to believe that it is raining) and the obligations undertaken (I have to adopt the principle of the sanctity of human life) without denying them (which would obviously be irrational).[24] Second, reasons can be more than one; indeed, I need at least one motivator, but in some cases there are many, and these reasons often conflict with one another; it then becomes necessary to appraise their relative weight in order to arrive at the prior intention and the intention-in-action.

In abductive reasoning, this kind of appraisal is linked to evaluating various inferred explanatory hypotheses/reasons, and, of course, it varies depending on the concrete cognitive and/or epistemological situation. In the section "Abductive Reasoning and Epistemic Mediators," below, I will illustrate that epistemologically using abduction as an inference to the best explanation simply requires evaluating competing hypotheses (which express competing "reasons" in the ethical case). The best total reason would be the one that creates prior intention and intention-in-action.

What criteria can we adopt to choose the reason(s) that will become the motivator(s)? In the last section of the previous chapter, I discussed the concept of "coherence" as illustrated by Paul Thagard,[25] in which ethical deliberation is seen as involving conflicting reasons (deductive, explanatory, deliberative, analogical) that can be appraised by testing their relative "coherence." This "coherence view" is terrifically interesting because it reveals the multidimensional character of ethical deliberations. The criteria for choosing the most coherent "reason/motivator" represent a possible abstract cognitive reconstruction of an ideal of "rationality" in moral decision making, but they can also describe the behavior of real human beings. As I have already stressed, human beings usually take into account just a fraction of all possible relevant knowledge when

[24] Searle, 2001, p. 115.
[25] Thagard, 2000.

performing ethical judgments. For example, when making judgments, it is common for utilitarians to employ only what Thagard calls "deliberative" coherence, or for Kantians to privilege principles over consequences. Psychological resources are limited for any agent, so it is difficult to mentally process all levels of ethical knowledge simultaneously in an attempt to calculate and maximize the overall coherence of the competing moral options. The "coherence" model accounts for these "real" cases of human moral reasoning by showing that they fit only "local areas" of the coherence framework: in general, real human beings come to immediate conclusions by considering only one moral level (for instance, the "consequentialist" one) and disregarding the possible change in coherence that could result from considering other levels (for instance, the "Kantian" one).

Searle interprets rationality in decision making naturalistically: "Rationality is a biological phenomenon. Rationality in action is that feature which enables organisms, with brains big and complex enough to have conscious selves, to coordinate their intentional contents, so as to produce better actions than would be produced by random behavior, instinct, tropism, or acting on impulse."[26] I agree, but I would add that rationality is a product of a hybrid organism. This notion obviously derives from the fact that even the external tools and models that we use in decision making – an externalized obligation, a computational aid, and even Thagard's "coherence" model described earlier – are products of biological human beings, while at the same time these tools constitutively affect human beings, who are, as we already know, highly "hybridized."[27]

Abduction in Practical Reasoning

Searle considers "bizarre" – and I strongly agree with him – that feature of our intellectual tradition according to which a true statement that describes how things are in the world can never imply a statement about how they ought to be. In reality, to take a simple example, to say that something is true is already to say that you ought to believe it – that is, other things being equal, that you ought not to deny it. Also, logical consequence can be easily mapped to the commitments of belief. Given the fact that logical inferences preserve truth, "The notion of a valid inference is such that, if p can be validly inferred from q, then anyone who

[26] Searle, 2001, p. 142.

[27] Cf. chapter 3, of this book. Searle (2001) calls the two aspects of performing an action (for instance, to fulfill an obligation) "effectors" and "constitutors." An obligation to another person is an example. I know I own you some money: "I can drive to your house" and "give you the money" – effector and constitutor.

asserts *p* ought not deny *q*, that anyone who is committed to *p* ought to recognize its commitment to *q*."[28] This means that normativity is more widespread than expected.

Certainly, theoretical reasoning can be seen as a kind of practical reasoning where deciding what beliefs to accept or reject is a special case of deciding what to do. The reason it is difficult to "deductively" grasp practical reasoning is related to the intrinsic multiplicity of possible reasons and to the fact that we can hold two or more inconsistent reasons at the same time.[29] The following example illustrates how practical contexts are refractory to logical modeling. Given the fact that I consider it a duty to do *p* and that I also feel committed not to do *p*, we cannot infer that I am committed to do (*p* and not *p*). I am a physician committed to not killing a patient who is in a coma, but at the same time my compassion for the patient commits me to the opposite duty. This does not mean that I want to preserve the life of the patient and, at the same time, that I want to kill him – that would lead to an inconsistent moral duty. All of this represents an unwelcome consequence of the fact that commitment to a duty is not closed under conjunction.[30]

In practical reasoning, we are always faced with desires, obligations, duties, commitments, needs, requirements, and so on that are at odds with one another. Moreover, even if I consider it a duty to do *p* and I believe that (if *p* then *q*), I am not committed to do *q* as a duty: I may be committed to killing a patient who is in a coma and at the same time believe that this act will cause pain for his friends, but I am not committed to causing this pain. *Modus ponens* does not work for the duty/belief mixture.[31]

These examples illustrate the difficulties that arise when classical logic meets practical reasoning. They further stress the importance I attribute to abductive explanatory inferences in practical settings, where creating, selecting, and appraising hypotheses are central functions. In the following sections, I will address the problem of abductive reasoning in science from a methodological perspective. The discussion can serve as a kind of appendix to further clarify many of the concepts I have explored in the book: creative, selective, and manipulative abduction; model-based reasoning; epistemic and moral mediators; explanatory reasoning; inference to the explanation, and so on.

28 Ibid., p. 148.

29 Searle "reluctantly" declares that it is impossible to construct a formal logic of practical reasoning "adequate to the facts of the philosophical psychology" (2001, p. 250). I think that many types of nonstandard logic (deontic, nonmonotonic, dynamic, ampliative, adaptive, etc.) reveal interesting aspects of practical reasoning by addressing the problem of the defeasibility of reasons and of their selection and evaluation.

30 Ibid., p. 250.

31 Ibid., pp. 254–255.

ABDUCTIVE REASONING AND EPISTEMIC MEDIATORS

This third part of the chapter is a kind of appendix that uses epistemology and cognitive science as a way to examine abductive reasoning's role in scientific settings. I will illustrate in detail the concepts of non-monotonicity of reasoning, model-based and manipulative abduction, and epistemic mediators as they relate to scientific thinking, so that readers may more fully understand the cognitive and epistemological ideas used in the book for studying issues of ethical reasoning and knowledge.

As we saw in the previous chapter, ethical deliberation, as a form of practical reasoning, is in some ways similar to hypothetical explanatory reasoning (selection and creation of hypotheses, inference to the best explanation) that occurs during abductive processes in scientific and diagnostic settings. Of course, instead of explanations that account for data, ethical cases have reasons that support conclusions, but moral reasons can still play an important explanatory role in those cases. In summary, building on some earlier segments of the book, the following sections will examine abduction as a valuable form of explanatory reasoning, explore model-based reasoning and epistemic mediators, and clarify the meaning of those concepts. The epistemic mediator, an idea I introduced in a previous book,[32] has suggested to me that moral reasoning also features some "external" structures that play an important ethical role. This has led me to individuate the concept of "moral mediator" that I described in the previous chapter.

What Is Abduction?

More than a hundred years ago, the great American philosopher Charles Sanders Peirce coined the term "abduction" to describe inference that involves generating and evaluating explanatory hypotheses. For example, Peirce said that mathematical and geometrical reasoning "consists in constructing a diagram according to a general precept, in observing certain relations between parts of that diagram not explicitly required by the precept,"[33] showing that these relations will hold for all such diagrams, and formulating this conclusion in general terms. What is abduction?

Many reasoning conclusions that do not proceed in a deductive manner are *abductive*. For instance, a shattered drinking glass on the floor is an *anomaly* that needs to be explained, and we might attribute it to a sudden strong gust of wind shortly before: this is certainly not a deductive

[32] Magnani, 2001a.

[33] That is, a kind of definition that prescribes "what you are to *do* in order to gain perceptual acquaintance with the object of the world." Peirce (*CP*), 2.33.

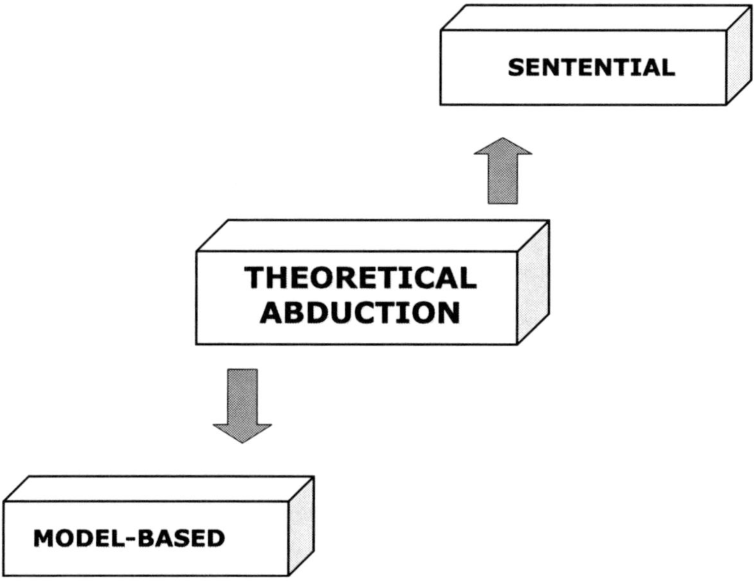

FIGURE 7.1. Theoretical abduction.

consequence of the glass being broken, for a cat may well have been responsible. Hence, what I call *theoretical* abduction[34] (Figure 7.1) is the process of *inferring* certain facts, laws, and hypotheses that render some sentences plausible, that *explain* or *discover* some (eventually new) phenomenon or observation; it is the process of reasoning by which explanatory hypotheses are formed and evaluated.

First, it is necessary to distinguish among abduction, induction, and deduction and to stress the significance of abduction as we analyze the problem-solving process. I think the example of diagnostic reasoning is an excellent way to introduce abduction. I have developed, with others, an epistemological model of medical reasoning called the Select and Test Model (ST-MODEL) that conforms to Peirce's stages of scientific inquiry: hypothesis generation, deduction (prediction), and induction.[35] The type of inference called abduction was also studied by Aristotelian syllogistics as a form of ἀπαγωγή and was explored later by medieval reworkers of syllogism. A hundred years ago, Peirce essentially interpreted abduction as an "inferential" (his particular use of the adjective "inferential" will be illustrated later) *creative process* of generating a new hypothesis. Abduction and induction, viewed together as processes for the production and

[34] Magnani, 1999 and 2001a.
[35] Lanzola et al., 1990; Ramoni et al., 1992; Magnani, 1992; Stefanelli and Ramoni, 1992.

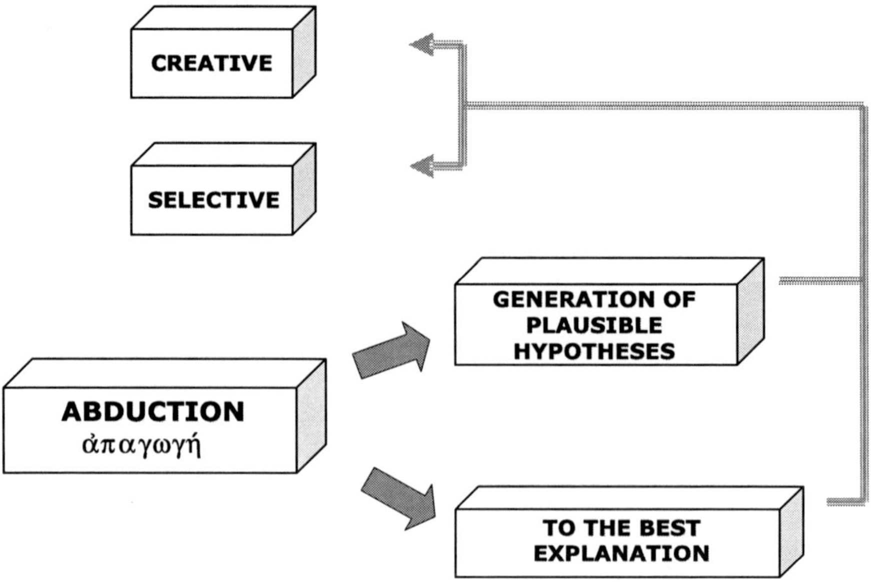

FIGURE 7.2. Creative and selective abduction.

generation of new hypotheses, are sometimes called "reduction," that is, again, ἀπαγωγή. As Łukasiewicz makes clear, "Reasoning which starts from reasons and looks for consequences is called *deduction*; that which starts from consequence and looks for reasons is called *reduction*."[36] The celebrated example given by Peirce is Kepler's conclusion that the orbit of Mars must be an ellipse.

There are two main epistemological senses of the word "abduction":[37] (1) abduction that generates only "plausible" hypotheses (*selective* or *creative* abduction) and (2) abduction that is considered *inference to the best explanation*, which also evaluates hypotheses (see Figure 7.2). It is clear that the two meanings are related to the distinction between hypothesis generation and hypothesis evaluation; abduction is the process of generating explanatory hypotheses, and induction matches the hypothetico-deductive method of hypothesis testing (first meaning). However, we have to remember (as we have already stressed) that sometimes in the literature (and also in Peirce's texts) the word "abduction" also refers to the whole cycle, that is, abduction is regarded as inference to the best explanation (second meaning).

[36] Łukasiewicz, 1970.
[37] Magnani, 1988, 1991.

In the chapter 6 section "Expected Consequences and Incomplete Information," I said that ethical reasoning is structurally similar to the reasoning that occurs in diagnosis. Diagnosis is a kind of *selective* abduction: to illustrate from the field of medical knowledge, the discovery of a new disease and its manifestations can be considered the result of a *creative* abductive inference. This is irrelevant in medical diagnosis, where the task is rather to *select* from an encyclopedia of pre-stored diagnostic entities.[38]

In the syllogistic view advocated by Peirce, which regards abduction as inference to the best explanation, one might require the final explanation to be the most "plausible" one. *Induction* in its widest sense is an ampliative process of the generalization of knowledge. While Peirce identifies various types of the process, all have in common the ability to compare individual statements:[39] using induction, it is possible to synthesize individual statements into general laws – inductive generalizations – in a defeasible way, but it is also possible to confirm or discount hypotheses. Following Peirce, I am referring here to the latter type of induction, that which reduces the uncertainty of established hypotheses by comparing their consequences to observed facts.

Deduction, on the other hand, is an inference that refers to a logical implication. In deduction, as opposed to abduction and induction, the truth of the conclusion is guaranteed by the truth of the premises on which it is based, so deduction involves so-called nondefeasible arguments. In order to illustrate these distinctions, it is useful to give a simple example of diagnostic reasoning in *syllogistic terms*, as Peirce originally did:[40]

1. If a patient is affected by pneumonia, his or her level of white blood cells is elevated.
2. John is affected by pneumonia.
3. John's level of white blood cells is elevated.

(This syllogism is known as Barbara).

By deduction, we can infer (3) from (1) and (2). Two other syllogisms can be obtained from Barbara if we exchange the conclusion (or Result, in Peircian terms) with either the major premise (the Rule) or the minor premise (the Case): by induction, we can go from a finite set of facts, like (2) and (3), to a universally quantified generalization – also called a categorical inductive generalization – like the piece of hematological knowledge represented by (1). Starting from knowing – selecting – (1)

[38] Ramoni et al., 1992.
[39] Peirce, 1955.
[40] See also Lycan, 1988.

and "observing" (3), we can infer (2) by performing a selective abduction. The abductive inference rule corresponds to the well-known fallacy called affirming the consequent (simplified to the propositional case):

$$\frac{\begin{array}{c}\varphi \rightarrow \psi \\ \psi\end{array}}{\varphi}$$

Thus, *selective abduction* involves a preliminary guess that generates plausible diagnostic hypotheses, which is then followed by deduction to explore their consequences and by induction to test them using available patient data. This process will either (1) increase the likelihood of one hypothesis by noting evidence for it rather than for competing hypotheses or (2) refute all but one hypothesis.

If during this first cycle new information emerges, hypotheses not previously considered can be evaluated, and a second cycle begins. In this case, the *nonmonotonic* character of abductive reasoning becomes evident, which arises from the logical unsoundness of the inference rule: it draws defeasible conclusions from incomplete information. A logical system is monotonic if the function *Theo* that relates every set of wffs to the set of their theorems holds the following property: for every set of premises S and for every set of premises S', $S \subseteq S'$ implies $Theo(S) \subseteq Theo(S')$.[41] Following this deductive nonmonotonic view of abduction, we can stress the fact that in actual abductive medical reasoning, when we increase the number of symptoms and the amount of patient data (premises), we are compelled to abandon previously derived plausible diagnostic hypotheses (theorems), as already illustrated by the ST-MODEL. All recent logical ("deductive") accounts concerning abduction have pointed out that it is a form of nonmonotonic reasoning. It is important to allow the guessing of explanations in order to discount old hypotheses and allow the tentative adoption of new ones, when new information about the situation makes them no longer the best.

We need a methodological criterion of justification, however: an abduced hypothesis that explains a particular puzzling fact should not automatically be accepted, because other worthy explanations are possible; a hypothesis that explains a certain number of facts is not guaranteed to be true. The combinatorial explosion of alternatives that has to be considered makes the task of finding the best explanation very costly. Peirce thinks that abduction has to be *explanatory* but also capable of experimental *verification* (that is, of being evaluated inductively; cf. the model presented

[41] Cf. Ginsberg, 1987; Lukaszewicz, 1990; Magnani and Gennari, 1997. On moral arguments that have a deductively valid or fallacious form (in the sense of classical logic), see Thomson, 1999, pp. 333–44. Recent research in the area of so-called practical reasoning, which relates to figuring out what to do, is illustrated in Millgram, 2001.

earlier), and *economic* (this includes the cost of verifying the hypothesis, its basic value, and other factors). Consequently, in order to achieve the best explanation, it is necessary to have or establish a set of criteria for evaluating the competing explanatory hypotheses reached by creative or selective abduction. Evaluation has a multidimensional and comparative character.

Consilience can measure how much a hypothesis explains and consequently can determine whether one hypothesis is more closely linked to the evidence (or data) than another: thus, consilience is a form of corroboration. One hypothesis is considered more consilient than another if it explains more "important" (as opposed to "trivial") data than the others do. In inferring the best explanation, the aim is not the sheer amount of data explained, but its relative significance.[42] The assessment of relative importance presupposes that an inquirer has a wealth of background knowledge about topics that concern the data. The evaluation is strongly influenced by Ockham's razor: *simplicity* too can be highly relevant when discriminating among competing explanatory hypotheses – for example, when dealing with the problem of the level of conceptual complexity of hypotheses when their consiliences are equal. Explanatory criteria are needed because the rejection of a hypothesis requires demonstrating that a competing hypothesis provides a better explanation.

Many approaches to finding such criteria have been proposed. For example, the so-called theory of *explanatory coherence* introduces seven principles that reflect the types of plausibility involved in accepting new scientific hypotheses and theories.[43] John Josephson has stressed that to evaluate abductive reasoning, one must ask the following questions: (1) How does a hypothesis surpass its alternatives? (2) How is the hypothesis good in itself? (3) How accurate is the data on which the hypothesis is based? and (4) How thorough was the search for alternative explanations?[44]

Finally, many attempts have been made to model abduction with tools that illustrate both its computational properties and its relationships with different forms of deductive reasoning. Some of the formal models of abductive reasoning are based on the theory of the *epistemic state* of an agent,[45] in which an individual's epistemic state is viewed as a consistent set of beliefs that can expand and contract (*belief revision framework*). This kind of *sentential framework* deals exclusively with selective abduction

[42] Thagard, 1988.
[43] Thagard, 1989, 1992.
[44] Josephson, 1998.
[45] Boutilier and Becher, 1995.

(diagnostic reasoning) and relates to the idea of preserving *consistency*.[46]

If we want a framework for analyzing Peirce's interesting examples of diagrammatic reasoning in geometry described earlier, we need not limit ourselves to the *sentential* view of theoretical abduction; we should consider a broader *inferential* approach that encompasses both sentential and what I have called *model-based* elements of creative abduction.

Thinking Through Drawing: Model-Based Abduction

Model-based reasoning is an important part of ethical reasoning, as I illustrated in the section "Model-Based Moral Reasoning" in the previous chapter. In scientific settings, the term "model-based reasoning" indicates the construction and manipulation of various representations, not only sentential and/or formal, but also mental and/or related to external mediators. Obvious examples of model-based reasoning are constructing and manipulating visual representations,[47] conducting thought experiments,[48] and engaging in analogical reasoning,[49] which occurs when models are built at the intersection of some domain and a new, ill-known domain. Nancy Nersessian has demonstrated that throughout the history of science, scholars and researchers have consistently used imagery and analogy to transform vague notions into scientifically viable conceptualizations. Her analysis deals with Faraday's and Maxwell's use of imagery and analogy in constructing the concept of a field.[50]

We have already seen that for Peirce all thinking is in signs, and that signs can be icons, indices, or symbols. Moreover, all *inference* is a form of sign activity, where the word *sign* includes "feeling, image, conception, and other representation."[51] For example, geometrical construction in elementary geometry is clearly a kind of model-based reasoning. In science, model-based reasoning acquires its peculiar creative relevance when embedded in abductive processes, such that we can individuate a *model-based abduction*.[52]

For Peirce, a Kantian key word is *synthesis*, a process by which the intellect constitutes and reconciles all the material delivered by the senses.[53] Kant did not consider synthesis to be a form of *inference*, but, notwithstanding the obvious differences, synthesis can be related to the

46 On the difference between "human" and "logical" agents in abductive reasoning, cf. Magnani, 2005.

47 Kosslyn and Koenig, 1992.

48 Brown, 1991.

49 Holyoak and Thagard, 1995.

50 Nersessian, 1995a and 1995b.

51 Peirce (*CP*), 5.283.

52 Magnani, 1999.

53 Anderson, 1986.

Peircian concept of inference and, consequently, to the concept of abduction. After all, when describing the ways in which the intellect unifies and constitutes phenomena through imagination, Kant himself uses the term *rule*: "Thus we think of a triangle as an object, in that we are conscious of the combination of the straight lines according to a rule by which such an intuition can always be represented."[54] He also employs the term *procedure* when he writes that "[t]his representation of a universal procedure of imagination in providing an image for a concept, I entitle the schema of this concept."[55] We know that rules and procedures are the central features of the modern concept of inference. Moreover, according to Peirce, the central question of philosophy is "how synthetical reasoning is possible.... This is the lock upon the door of philosophy."[56] He goes on to say that the mind tends to unify the many aspects of a phenomenon: "the function of conception is to reduce the manifold of sensuous impressions to unity."[57]

Most of these forms of constitution of phenomena are creative and, moreover, characterized in a model-based way. In the previous chapter, I described perception and emotion as forms of model-based abduction and discussed the role of model-based reasoning in various cases of ethical reasoning. In another interesting example of model-based abduction, Peirce observes that the perception of tone occurs after the mind has noted the rate of a sound wave's vibration, but that a tone can be identified only after a person has heard several sound impulses and has judged their frequency. Consequently, the sensation of pitch is made possible by previous experiences and cognitions stored in memory, so that a single sound wave would not produce a tone.

We can revisit Peirce's observation about touch: "A man can distinguish different textures of cloth by feeling: but not immediately, for he requires to move fingers over the cloth, which shows that he is obliged to compare sensations of one instant with those of another."[58] This idea teaches us that abductive processes also have crucial extratheoretical features and that abductive reasoning can involve manipulations of external objects. When manipulative aspects of external models prevail, as when we draw diagrams on the blackboard, we face what I call *manipulative* abduction (or action-based abduction)[59] (see the following section).

What happens when abductive reasoning in science is strongly related to extratheoretical actions and manipulations of "external" objects? What

[54] Kant, 1929, A105, p. 135.
[55] Kant, 1929, A140/B179–180, p. 182.
[56] Peirce (*CP*), 5.348.
[57] Ibid., 1.545.
[58] Ibid., 5.221.
[59] Magnani, 2000.

is the result of "action-based" abduction conducted through *external models*, as is the case when we use external geometrical diagrams? What occurs when thinking is "through doing," as in Peirce's example of distinguishing cloth by feeling? To answer these questions, I have delineated the first features of what I call *manipulative abduction* by showing how we can find, in scientific and everyday reasoning, methods of constructivity based on external models and actions.

Thinking Through Doing: Manipulative Abduction

The problem of the incommensurability of meaning has distracted epistemologists from procedural, extra-sentential and extra-theoretical aspects of scientific practice. Since Thomas Kuhn, the problems of language translation and conceptual creativity have dominated the theory of meaning.[60] *Manipulative* abduction (Figure 7.3) happens when we are thinking *through* doing and not only, in a pragmatic sense, *about* doing. So in manipulative abduction, experiments do more than predict outcomes or establish new scientific laws through their results: manipulative abduction refers to extra-theoretical behavior that allows us to communicate new experiences and integrate them into previously existing systems of experimental and linguistic (theoretical) practice.

We have already said that this kind of extra-theoretical cognitive behavior occurs in everyday situations in which people repeat common but complex tasks without really thinking about them.[61] In this kind of *action-based* abduction, the suggested hypotheses are inherently ambiguous until articulated as configurations of real or imagined entities (images, models, or concrete apparatus and instruments). In these cases, we can discriminate between possibilities only through experimentation: they are articulated behaviorally and concretely by manipulations and then, increasingly, in words and pictures.

Manipulative reasoning is used frequently in science. As indicated in the previous chapter (in the section "Being Moral through Doing: Taking Care"), David Gooding refers to this kind of concrete manipulative reasoning when he illustrates the scientific role of the so-called construals that embody tacit inferences in visual and tactile procedures that often involve machines or other apparatus.[62] Construals belong to the preverbal context of ostensive operations that are practical, situational, and often made with the help of words, visualizations, or concrete artifacts. This embodied expertise allows one to adeptly manipulate objects in a highly constrained experimental environment and gives rise to abductive

[60] Kuhn, 1970.
[61] Hutchins, 1995.
[62] Gooding, 1990.

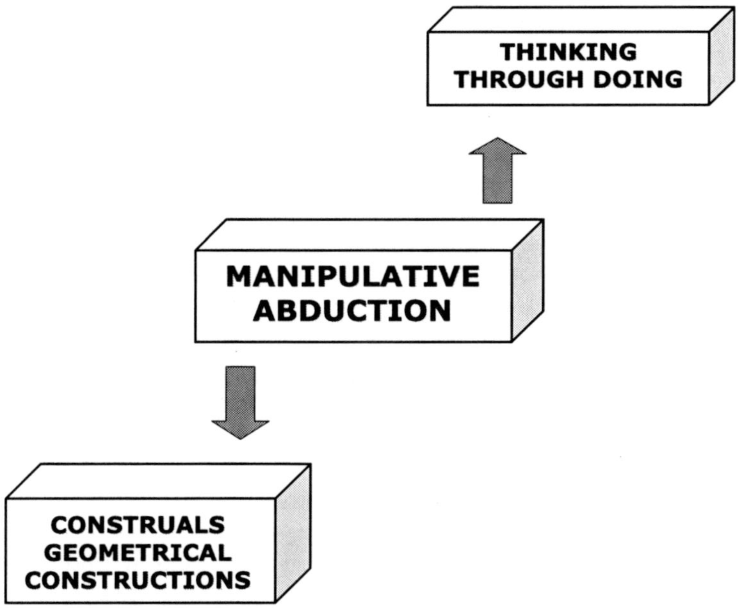

FIGURE 7.3. Manipulative abduction.

movements that involve the strategic application of old and new (non-conceptual) *templates* of behavior connected, in some cases, to extra-rational components.

The hypothetical character of construals is clear: they can be developed to examine further opportunities or they can be discarded; they provisionally organize experience in a creative manner, and in turn, some of them become more theoretical *interpretations* of experience that are gradually stabilized through established observational practices. Step by step, an initially practice-laden interpretation becomes a more theory-oriented mode of understanding (narrative, visual, diagrammatic, symbolic, conceptual, simulative) that more closely resembles theoretical abduction. When the reference is stabilized, any incommensurability with other established observations becomes evident. But only the construals of certain phenomena can be shared by supporters of rival theories. Gooding shows how Davy and Faraday could see the same electromagnetic attraction and repulsion at work in the phenomena they respectively produced; their discourse and practice regarding their construals of phenomena clearly demonstrate that in some cases they did not inhabit different, incommensurable worlds.

Also of extreme interest are Gooding's studies on manipulating visual representations, as they show how scientists generate new cognitive and social resources in order to reduce cognitive demands and to expand the scope and application of representation. The study identifies a common set of visual transformations underlying the methods and imagery of different scientific fields. Transformations of visual representations are also analyzed in ways that achieve other kinds of matches: that "between a representation and the cognitive demands of a task (such as pattern matching or mental rotation) or between an emerging representation and the social need to communicate and negotiate new meanings in the context of culturally embedded conventions."[63] Relating visualization to construals, Gooding says: "Visualization works in conjunction with investigative actions to produce a phenomenology of interpretive images, objects, and utterances," and he acknowledges my account of this process in terms of manipulative abduction.[64]

Also related to manipulative abduction is the study of technological evolution in the case of the invention of the telephone. Michael Gorman evaluates two innovators, Alexander Graham Bell and Elisha Gray, as he shows how they created tone-transmitting devices.[65] He indicates how these researchers were involved in a kind of conversation among internal mental models, mechanical representations, and their manipulations of them; as he does so, Gorman takes advantage of the heuristics that anchor the cognitive process of invention to the hybrid interplay between individual and external representations and devices:

> Cognition clearly is both embodied in brain, hands, and eyes, and also distributed among various technologies and shared across groups. ... in order to compare two inventors, one needs to understand their mental models and mechanical representations. This kind of understanding is possible only because so much of cognition is in the world.[66]

The theory of manipulative abduction also supports, for example, Thagard's statement that, the school of eighteenth-century scientists who believed phlogiston was required for combustion and those who thought oxygen was the critical component could recognize each other's experiments.[67] Thagard's assertion is exhibited as an indispensable requisite for his coherence-based epistemological and computational theory of comparability at the level of intertheoretic relations and for the whole problem of creative abductive reasoning to the best explanation.

63 Gooding, 2004, p. 552.
64 Gooding, 2005, p. 197.
65 Gorman, 1997. See also Carlson and Gorman, 1990.
66 Cf. Gorman, 1997, pp. 589, 612. I have illustrated the related aspects of what he calls "trading zones" in chapter 4.
67 Thagard, 1992.

Gooding introduces the so-called *experimental maps*,[68] which are two-dimensional epistemological tools that illustrate the conjecturing (abductive) role of actions from which scientists "talk and think" about the world. They are particularly useful in that they call attention to the interaction of hand, eye, and mind inside the actual four-dimensional scientific cognitive process. The various procedures for manipulating objects, instruments, and experiences will in turn be reinterpreted as procedures for manipulating concepts, models, propositions, and formalisms.

Scientists' activity in material environments results in rich perceptual experiences that, generally speaking, are described as visual experiences using constructive and hypothesizing narratives. Moreover, the experience is constructed, reconstructed, and distributed across a social network of different scientists by means of construals[69] that reconcile conceptual conflicts. As I have said, construals constitute a provisional creative organization of experience: when they become hypothetical interpretations of experience – that is, more theory-oriented – their reference is gradually stabilized through established observational practices that also exhibit a cumulative character. It is in this way that scientists are able to communicate the new and unexpected information acquired by experiment and action.

To illustrate this progression from manipulations, to narratives, to possible theoretical models (visual, diagrammatic, symbolic, mathematical), we need to consider some observational techniques used by Faraday, Davy, and Biot concerning Oersted's experiment on electromagnetism. They were able to create consensus because their conjectural representations successfully resolved phenomena into stable perceptual experiences. Some of these narratives are very interesting. For example, Faraday observes that "it is easy to see how any individual part of the wire may be made attractive or repulsive of either pole of the magnetic needle by mere change of position. . . . I have been more earnest in my endeavors to explain this simple but important point of position, because I have met with a great number of persons who have found it difficult to comprehend." Davy comments: "It was perfectly evident from these experiments, that as many polar arrangements may be formed as a chord can be drawn in circles surroundings the wire." Expressions in the narratives like "easy to see" and "it was perfectly evident" are textual indicators of the stability of the forthcoming interpretations. Biot, in his turn, provides a three-dimensional representation of the effect by giving a verbal account that enables us

[68] Circles denote mentally represented concepts that can be communicated; squares denote things in the material world (bits of apparatus, observable phenomena) that can be manipulated; and lines denote actions.

[69] Cf. Minski, 1985, and Thagard, 1997a.

to visualize the setup – "suppose that a conjunctive wire is extended horizontally from north to south, in the very direction of the magnetic direction in which the needle reposed, and let the north extremity be attached to the copper pole of the trough, the other being fixed to the zinc pole ..." – and then describes what will happen by illustrating a sequence of steps in a geometrical way:

> Imagine also that the person who makes the experiment looks northward, and consequently towards the copper or negative pole. In this position of things, when the wire is placed above the needles, the north pole of the magnet moves towards the west; when the wire is placed underneath, the north pole moves towards the east; and if we carry the wire to the right or the left, the needle has no longer any lateral deviation, but it loses its horizontality. If the wire be placed to the right hand, the north pole rises; to the left, its north pole dips.[70]

It is clear that "seeing" interesting things in this experimental context depends on a manipulation's ability to obtain correct information and suggest new interpretations (for example, a simple mathematical form) of electromagnetic natural phenomena, a process that occurs on the theoretical side of abduction. Step by step, we arrive at Faraday's account in terms of magnetic lines and curves.

Establishing a list of standard behaviors that illustrate scientific manipulative abduction is difficult. As we have seen, expertly manipulating objects in a highly constrained experimental environment requires both old and new behavioral *templates*, for construals are very conjectural and do not immediately provide explanations: these templates are hypotheses of behavior, either newly created or drawn from those already stored in the scientist's mind-body system, that abductively enable a kind of epistemic "doing." Hence, some templates can be *selected* from an existing collection, while others must be *created* on the spot in order to perform the most interesting cognitive acts of manipulative abduction.

Moreover, I think a better awareness of manipulative abduction in scientific settings could improve our understanding of induction and how it differs from abduction: manipulative abduction could be considered a basis for further inductive generalizations, and different construals can give rise to different inductive generalizations.

In the previous chapter (in the section "Templates of Moral Doing"), I outlined characteristics of templates used in ethical reasoning, and these correspond to behavioral templates employed in scientific work. There

[70] The quotations are from Faraday, 1821–22, p. 199; Davy, 1821, p. 16; and Biot, 1821, pp. 282–283, cited in Gooding, 1990, pp. 35–37.

are at least four common features of tacit templates that enable us to manipulate things and experiments in science:

1. *Sensibility to curious or anomalous aspects of the phenomenon.* Manipulations must be able to expose potential inconsistencies in the received knowledge. Consider, for example, that Oersted's report on his well-known experiment about electromagnetism focuses on anomalies that did not depend on any particular theory of the nature of electricity and magnetism, and that Ampère's construal of his electromagnetism experiment, in which he used an artifactual apparatus to produce a static equilibrium of a suspended helix, clearly shows the role of the "unexpected."
2. *Preliminary sensibility to the dynamic character of the phenomenon* rather than to entities and their properties. A common aim of manipulations is to reorganize the dynamic sequence of events into a static spatial order that should allow a bird's-eye view, either narrative or visual-diagrammatic, of the experiment.
3. *The use of artificial apparatus in scientific manipulations* to identify possibly stable and repeatable sources of information about hidden knowledge and constraints. Davy's artifactual tower of needles, for instance, showed that magnetization was related to orientation and does not require physical contact. Of course, information obtained in this way is not artificially constructed: the fact that phenomena are made and manipulated does not mean that they are idealistically and subjectively determined.
4. *Various contingent ways of epistemic acting*, which might include *looking* from different perspectives; *checking* the various kinds of information available; *comparing* subsequent events; *choosing*, *discarding*, and *imaging* further manipulations; and *reordering* and *changing relationships* in the world by implicitly *evaluating* the usefulness of a new order.

In summary, manipulation is devoted to building external *epistemic mediators* that function as a vital new source of information and knowledge. Therefore, we can say that manipulative abduction redistributes epistemic and cognitive effort in a way that helps us to manage objects and information that cannot be immediately represented or found internally. For example, it is difficult to preserve precise spatial relationships using mental imagery, especially when one set of them has to be moved relative to another.[71]

[71] I derived the expression "epistemic mediator" from the cognitive anthropologist Edwin Hutchins, who coined the expression "mediating structure" to refer to various external tools that can facilitate cognitive acts of navigating in both modern and "primitive" settings. Any written procedure is a simple example of a "mediating structure" with possible cognitive aims: "Language, cultural knowledge, mental models, arithmetic procedure, and rules of logic are all mediating structures, too. So are traffic lights,

If we see scientific discovery as an opportunity to integrate information from simultaneous constraints and, subsequently, to produce explanations for those constraints, then manipulative abduction can elicit hidden constraints as it builds external experimental structures. So well-built external structures (Biot's construals, for example) and the new knowledge they contain will be projected onto internal structures like models and symbolic frameworks – all of which results in constructive theoretical abduction. When we superimpose the internal and the external, we generate a provocative interplay between manipulative and theoretical abduction, and these novel juxtapositions reveal new relationships and meanings. This interplay also shows us that, in fact, internal and external processes are part of the same epistemic ecology, an idea that Hutchins alludes to when he uses the phrase "cognitive ecology" when discussing internal and external cognitive navigational tools.[72]

Of course, many of the actions that are entertained to build such artifactual models are not tacit, but explicitly projected and planned. However, imagine the people who were the first to create some epistemic artifacts (for instance, Faraday and Biot): they created them simply and mainly "through doing" (by the creation of new construals through tacit templates) and surely not by following already well-established projects.

Picking Up Information

We can say that abduction is a complex *process* that works through *imagination*: it suggests a new direction of reasoning by *shaping* new ways of explaining (cf. the templates mentioned earlier). Imagination should not, however, be confused with intuition. Peirce describes abduction as a dynamic modeling process that fluctuates between states of doubt and states of belief. To assuage doubt and account for anomalies, the agent gathers information that relates to the "problem," to the agent's evolving understanding of the situation and its changing requirements. When I use the word "imagination" here, I am referring to this process of knowledge gathering and shaping, a process that Kant says is invisible and yet leads us to *see* things we would not otherwise have seen: it is "a blind but indispensable function of the soul, without which we should not have no knowledge whatsoever.[73] For example, scientific creativity, it is pretty

supermarket layouts, and the contexts we arrange for one another's behavior. Mediating structures can be embodied in artifacts, in ideas, in systems of social interactions." (Hutchins, 1995, pp. 290–291)

[72] Ibid., 1995, p. 114. More observations about manipulative abduction can be found in Morgan and Morrison, 1999, which deals with the mediating role of scientific models between theory and the "real world."

[73] Kant, 1929, A78/B103, p. 112.

obvious, involves seeing the world in a particular new way: scientific understanding permits us to see some aspects of reality in a particular way, and creativity relates to this capacity to shed new light.

We can further analyze this process using the *active perception* approach, a theory developed in the area of computer vision.[74] This approach seeks to understand cognitive systems in terms of their environmental *situatedness*: instead of being used to build a comprehensive inner model of its surroundings, the agent's perceptual capacities are simply used to obtain whatever specific pieces of information are necessary for its behavior in the world. The agent constantly "adjusts" its vantage point, updating and refining its procedures, in order to uncover a piece of information. This leads to the need for specifying how to examine and explore efficiently and to the need for "interpreting" an object of a certain type. It is a process of attentive and controlled perceptual exploration through which the agent is able to collect the necessary information: a purposefully moving through what is being examined, actively picking up information rather than passively transducing.[75]

As suggested, for instance, by Lederman and Klatzky, this view of perception may be applied to all sense modes: for example, it can easily be extended to the haptic mode.[76] Mere passive touch, in fact, tells us little, but by actively exploring an object with our hands we can find out a great deal. Our hands incorporate not only sensory transducers, but also specific groups of muscles that, under central control, move them in appropriate ways: lifting something tells us about its weight; running fingers around the contours provides information about its shape; rubbing it reveals, for instance, its texture, as already stressed by Peirce in the quotation reported earlier, when dealing with the hypothesizing activity of what I call manipulative abduction.[77]

Nigel Thomas suggests that we think of the fingers together with the neural structures that control them and the afferent signals they generate as a sort of knowledge-gathering (perceptual) *instrument*: a complex of physiological structures capable of active testing for some environmental property.[78] The study of manipulative abduction outlined earlier can benefit from this approach. For example, the role of particular epistemic mediators (optical diagrams) in nonstandard analysis has been studied, as well as their function in grasping and teaching abstract and difficult mathematical concepts.[79] In this case, the external models (mathematical

74 Thomas, 1999.
75 Cf. Gibson, 1979.
76 Lederman and Klatzky, 1990.
77 Peirce (*CP*), 5.221.
78 Thomas, 1999.
79 See Magnani and Dossena, 2002.

diagrams) do not provide all the available knowledge about a mathematical object, but compel the agent to engage in a continual epistemic dialogue between the diagrams themselves and its internal knowledge in order to enhance understanding of existing information or to facilitate the creation of new knowledge.

It is clear that human beings and other animals make frequent use of perceptual reasoning and kinesthetic abilities. We can catch a ball, cross a busy street, read a musical score, evaluate a shape by touch, or recognize a friend's partially obscured face, for example. Usually, the "computations" these tasks require do not occur at a conscious level. Mathematical reasoning uses verbal explanations, but it also involves nonlinguistic notational devices and models that require our perceptive and kinesthetic capacities. Geometrical constructions are a relatively simple example of this kind of extra-linguistic machinery that functions in a model-based, manipulative, and abductive way.

COGNITIVE AND EPISTEMIC MEDIATORS

Recent research, taking an ecological approach to analyzing and designing human-machine systems,[80] has shown how expert performers create an *external* model of task dynamics in everyday life that can be used in lieu of an internal model: Alex Kirlik says, "Knowingly or not, a child shaking a birthday present to guess its contents is dithering, a common human response to perceptually impoverished conditions."[81] Action is more than just a way to move the world to a desirable state – it also performs an epistemic role: people choose behaviors that not only simplify cognitive tasks but also compensate for incomplete information or the inability to act upon the world. *Epistemic action* can also be the result of latent constraints in the human-environment system. This additional constraint grants additional information: when a child shakes a birthday present, he "takes actions that will cause variables relating to the contents of the package to covary with perceptible acoustic and kinesthetic variables. Prior to shaking, there is no active constraint between these hidden and overt variables causing them to carry information about each other." Similarly, "one must put a rental car 'through its paces' because the constraints active during normal, more reserved driving do not produce the perceptual information necessary to specify the dynamics of the automobile when driven under more forceful conditions."[82]

[80] Kirlik, 1998.

[81] Ibid.

[82] Ibid. Kirlik also describes how a group of short-order cooks of varying levels of skill in Atlanta used external models of the dynamic structure of the grill surface to get new information that was otherwise inaccessible.

In related work, Powers has studied behavior both as the *control of perception* and as controlled by perception.[83] Flach and Warren use the term "active psychophysics" to illustrate that "the link between perception and action ... must be viewed as a dynamic coupling in which stimulation will be determined as a result of subject actions. It is not simply a two-way street, but a circle."[84] Kirsh describes situations (e.g., grocery bagging, salad preparation) in which people use action to simplify choice and perception and to reduce the demands for internal computation through the exploitation of spatial structuring.[85]

We know that theoretical abduction certainly illustrates much of what is important in abductive reasoning, especially the objective of selecting and creating a set of hypotheses (diagnoses, causes, reasons) that are able to dispense good (preferred) explanations of data (observations), but fails to account for many kinds of explanations occurring in science or in everyday reasoning when the exploitation of the environment is crucial. Manipulative abduction, on the other hand, reveals the role of action in many interesting situations: action provides otherwise unavailable information that enables the agent to solve problems through an abductive process of generating or selecting hypotheses.

Manipulative abductive reasoning – that is, action – can offer very interesting templates for use in everyday situations:

1. It *simplifies* the reasoning task and redistributes effort across time[86] when we "need to manipulate concrete things in order to understand structures which are otherwise too abstract,"[87] or when we are dealing with *redundant* and unmanageable information. Extremely interesting examples include not only Gooding's work on manipulating artifacts in experiments (construals) but also his studies, quoted earlier, on the manipulation of visual representations as facilitating cognitive processes like pattern matching and visual inference.[88]
2. Action can be useful in the face of *incomplete* or *inconsistent* information – not just from the "perceptual" point of view – and when one is hampered by a diminished capacity to act upon the world: action can be used to get additional data that will restore coherence and improve deficient knowledge.
3. As the *control of sense data*, action makes new kinds of stimulation possible by changing the position of our bodies and/or external

[83] Powers, 1973.
[84] Flach and Warren, 1995, p. 202.
[85] Kirsh, 1995.
[86] Hutchins, 1995.
[87] Piaget, 1974.
[88] Gooding, 2004.

objects and by exploiting various kinds of prostheses and technological instruments and interfaces. Examples of external prostheses are Galileo's telescope, and the surgical instruments that a physician uses in an exploratory operation to pin down the cause of a patient's pain – such tools give access to tactile and visual information that would otherwise be unavailable.

4. Action enables us to build *external artifactual models* of task mechanisms to supplement the corresponding internal ones so that the environment is adapted to the agent's needs.

As we saw in the previous chapter's discussion of moral mediators, natural phenomena can also serve as external artifactual models: the stars are not artifacts, but as a Micronesian navigator manipulates images of them, they acquire a structure that "becomes one of the most important structured representational media of the Micronesian system."[89] Such external artifactual models are parts of a memory system that crosses the boundary between person and environment; they are able, for example, to transform the tasks involved in allowing simple manipulations that promote further visual inferences at the level of model-based abduction.[90] I have illustrated the ethical side of these features in the previous chapter, in the section "Templates of Moral Doing."

Not all epistemic and cognitive mediators are preserved, saved, and improved, as in the case of the ones created by Galileo at the beginning of modern science. For example, in certain everyday, non-epistemological emergency situations some skillful mediators are constructed to face possible dangers, but because such events are rare, the mediators are not saved and stabilized. Hutchins describes the interesting case of the failure of a gyrocompass, an electrical device crucial for navigation, and how sailors created substitute cognitive mediators: the crew made additional computations, redistributed cognitive roles, and finally, instituted a new shared mediating artifact by shifting the division of labor – the so-called modular sum that makes the situation manageable.[91]

89 Hutchins, 1995, p. 172.

90 The study of mechanisms of manipulative abduction may also improve technological interfaces that provide restricted access to controlled systems, so that humans have to compensate by reasoning with and constructing internal models. New resources for actions deriving from improved interfaces, related to task-transforming representations, can contribute to overcoming these reasoning obstacles (Kirlik, 1998, and Kirsh, 2006). On the interplay between internal and external representations in the so-called multimodal abduction, cf. Magnani, 2006a. On Gibsonian "affordances" and abduction, cf. Magnani and Bardone, 2007.

91 Hutchins, 1995, pp. 317–351.

We must note that many external things that we normally consider epistemologically inert can be transformed into epistemic or cognitive mediators. Consider our bodies, for example: we can talk with ourselves, which is a sort of self-regulatory action; use our fingers for counting; and use external "tools" like writing, narratives, information obtained from other people,[92] concrete models and diagrams, and various artifacts. Another example might be the gestures that accompany speech – sometimes we use them while talking, sometimes before or after we say something. All this is to say that not every cognitive tool is inside the head and that external objects and structures are often useful epistemic devices. As we saw earlier, even natural objects can become complicated epistemic artifacts, as occurred when the Micronesian navigator's stars were inserted into cognitive manipulations (of seeing them) related to navigation.

I contend that using a methodological approach to rethink casuistry is a particularly useful exercise: it shows that judgments, especially those involving complicated cases, do not always derive from rigid, well-established, all-encompassing moral principles but often result from verbal argument that casuistically take into account particular circumstances, concrete aspects, precedents, and possible exceptions. This revaluation is compelling given the fact that inexorable technological advances will continue to create unexpected challenges with many particulars to explore; we cannot meet these challenges successfully using only traditional moral principles and orthodox ways of thinking.

Finally, I would like to reiterate the importance of the section "The Logical Structure of Reasons," in which my cognitive consideration of moral mediators and morality "through doing" reveals the distinction between "internal" and "external" reasons in ethical deliberation. "External reasons" refer to various externalities I have mentioned throughout the book, starting with my proposal to "respect people as things" and continuing through the discussion of "moral mediators." It is in this section that I explicitly work out the main theoretical aspects of my thesis about cognitive and epistemological traditions: when applied in new ways to the study of moral deliberation, they can help us to revitalize research in ethics and to confront impending technological consequences head on.

[92] The results of empirical research that show the importance of collaborative discovery in scientific creative abduction and in explanatory activities are given in Okada and Simon, 1997.

Afterword

In place of a formal conclusion, I offer here a sort of summary and a few final comments about what I see as the most important elements of *Morality in a Technological World*. My hope is to have established a compelling rationale for making a serious commitment to new knowledge, both scientific and ethical, for it is only through new knowledge that we can succeed in "respecting people as things" or, for that matter, "treating people as ends." The new knowledge I passionately endorse will supply us with the moral poise required to handle controversial technology now and in the future. With such knowledge, we will be better prepared to accept rather than fear the new types of hybrid people that will result from technology; ironically, by respecting people as things, we may yet construct Immanuel Kant's shining "kingdom of ends."

Those who are suspicious of biotechnologies often invoke Kant's dictum that we must not treat people as means, but that maxim alone, here in the increasingly complex twenty-first century, can sometimes seem too simplistic and general to do us much good in the face of cloning and other practices. But we can build on Kant's idea: by inverting his mandate – by *encouraging* people to treat others as things – we can begin to make peace with inevitable technological advances. Thinking of human beings as valuable "things" is a useful way to navigate the murky waters of many modern ethical problems related to technology, which are certain to become more rather than less complicated. This unavoidable fact requires us to treat knowledge as our duty if we wish to live as moral beings now and, especially, in the future, when "monsters" – those who are cloned, genetically enhanced, or somehow hybrid – may be as common as the mildly hybrid people who today wear contact lenses to enhance their vision.

The increasing hybridization of the human and the artificial, a process fueled, of course, by scientific knowledge, makes "respecting people as things" more and more relevant to our everyday lives; similarly, it demands greater understanding of this new hybrid human condition. It was by this intellectual route that I reached the counterintuitive conclusion that it may be preferable to treat people as things if we wish to improve their lives. If the part-thingness of human beings is widely acknowledged throughout society, perhaps politicians and others who claim to value human dignity but behave otherwise can learn to view the world differently: those leaders who freely lavish both economic and moral value on externalities may learn from those "things" and see how their worth can be transferred to people. I believe many changes are possible once we enter into the details of this "thingness," which already affects human beings and will affect them even more in the future.

That which is human continues to become more deeply and inextricably intertwined with the nonhuman, and this mixture blurs the established distinction between people and things: we are "folded" into nonhumans, so that we delegate cognition and action to external things that coexist with us. The knowledge we have achieved in the past – scientific, social, moral, and so on – has already transformed us into hybrid people that embody both natural and artifactual cognitive functions, a process that is hastened by the mediation of modern technology. The line between the biological self and the technological world is now extremely flexible, and this fact has to be acknowledged from both the epistemological and the ontological point of view in order to be fully appreciated.

I contend that as our ethical knowledge evolves and strengthens, so too will our free will, consciousness, and intentionality, some of the central components of human dignity. These human aspects are constantly jeopardized in unexpected ways by technological products, as illustrated in the first two chapters, but they are also threatened in some not-so-surprising ways: unethical actions by opportunistic criminals, well-meant actions that turn out to be mistakes, and political wrangling gone awry, for example. We are not always wholly cognizant of the fact that externalities can be hazardous to our dignity – as defined by the factors just mentioned – and it is this lapse of awareness that has underscored for me the intrinsic limitations of the so-called ethics of science and challenged me to construct a more useful ethical perspective on "knowledge." I consequently propose that we in the ethics community and scholars in other fields emphasize knowledge that facilitates not only human wealth and happiness but also human dignity, a goal that requires us to maintain, improve, and in some cases establish for the first time widespread accountability, free choice, and control over individual destinies. In so doing, we can transcend the baleful predictions of a future with half-human creatures and organ farms.

To embrace knowledge as duty is to commit to the future and take seriously the notion that our choices have great impact, not just on ourselves and our immediate families and friends, but on untold numbers of strangers, both present and future. New knowledge is needed to manage dangerous or unethical consequences that arise from technology unexpectedly, and the more wisely we interrogate and monitor today's choices, technological or otherwise, the less the people of tomorrow will need to sacrifice to ensure the continuation of humanity on Earth. Modern human behavior leaves indelible marks on the world in a way that was impossible a few generations ago, which means that we must tread more delicately and conscientiously. While still admirable goals, "neighbor ethics" like justice, honesty, and charity are not enough to offset the possible deleterious effects of certain actions; our technological capabilities have become so far-reaching, so profoundly pervasive – and will become more rather than less so – that collaborative initiatives by collectives are our only hope to significantly slow the damage human beings inflict on each other and on the environment. Our current activities may not destroy the Earth itself, but, over time, they may very well render growing portions of it uninhabitable for human beings.

We are obliged to acquire and implement knowledge, but it is *not* our duty to make all information and knowledge available to anyone who wants it. Recent strides in communications and computer technology, while beneficial in many ways, pose increasingly great risks to identity and cyberprivacy. If too much knowledge is incorporated into external artificial things, the "visibility" of people can become dangerously excessive, and they risk diminished privacy, decreased control, less protection against interference, and weakened cyberdemocracy. We must be vigilant about monitoring the moral effects of distributing knowledge, and we must learn to anticipate possible negative outcomes and unethical consequences of evolving information policies.

Of course, if knowledge is to be considered a duty, then we must also consider any related cognitive, logical, methodological, and epistemological problems along with issues regarding its dissemination. Here, in particular, is where our willingness to shed the shackles of traditional ethics becomes crucially important; fully understanding these concepts requires entirely new ethical commitments. The many issues needing attention reflect possible trends and suggest possibilities for future research: from the need for studying creativity and model-based and manipulative thinking in scientific and ethical reasoning to expanding the role of "moral mediators"; from the interplay between unexpressed and super-expressed forms of knowledge and knowledge management in information technology to the challenge of forming interesting rational ethical arguments. I also emphasize the need to better understand the "intrinsic value" of knowledge (and general information) in metaethics itself: knowledge is duty, but who owes it to whom? And what about the related right to knowledge (and information)? The lack of appropriate and situated knowledge contributes to a negative bias in concrete moral deliberations and poses an obstacle to responsible behavior. As a result, I also submit that we should examine the role of the new so-called knowledge communities, trading zones, ESEM capacities, and the status of humans beings as "knowledge carriers."

Indeed, my approach can be characterized in terms of a combination of ethics, epistemology, and cognitive science. I am convinced that moral concerns involve reasoning that bears important parallels to reasoning in the sciences and that these similarities can help us to address the problem of moral deliberation in cases and problems not anticipated by moral philosophers. Hence, by using disciplines not usually associated with ethics – like epistemology and cognitive science – I think it is possible to rethink and retool research on the ethics and philosophy of technology. I think readers will agree that it is difficult to increase knowledge about ethical reasoning when it comes to problems like creativity and non-monotonicity, as well as sentential, model-based, embodied, and distributed aspects, without considering recent findings in logic, epistemology, and the whole field of cognitive science; this is why I have frequently alluded to some components of my previous epistemological and cognitive research on the concept of abduction.

Most of the concepts in *Morality in a Technological World* can be further clarified using the two-pronged methodological approach of the "twin" chapters at the end of the book. Complementary strategies allow us to reorient and modernize philosophical discussions of many ethical issues in a way that avoids the formal treatments of traditional moral philosophy, which demand an unrealistic degree of competence, information, and (formal) reasoning power. Approaching moral reasoning in a novel way – through both epistemology and cognitive results – shows that we can appeal to scientific thinking and problem solving for practical reasoning models in the ethics and philosophy of technology: the source is especially appropriate in light of the fact that science and technology underlie many of the social and cultural changes that require an up-to-date approach to ethics.

This framework cognitively and methodologically delineates the new concept of "moral mediators," entities we can construct in order to bring about certain ethical effects. These mediators may exist as beings, objects, or structures; any of these may carry unintended ethical or, in the case of some previously discussed technological artifacts, unethical consequences. I believe that the potential for moral mediators and "respecting people as things" is remarkable: these ideas let us analyze the condition of modern people in entirely new ways, which is critically important because modern forces like globalization have diluted and distorted the old ways, clouding our vision to the extent that we cannot "see" the value of millions of human beings.

More attention must be paid to our thinking patterns and habits, and abduction, the epistemological concept that most clearly illustrates hypothetical reasoning, both elucidates the moral processes of "inferring reasons" and reveals the differences between "internal" and "external" reasons in ethical deliberation. I think ethicists sometimes mistrusted practical reasoning because it is hard to grasp it "deductively." At least in classical logic, this difficulty is mainly due to the intrinsic multiplicity of possible reasons and to the fact that in practical reasoning we often hold two or more inconsistent reasons at the same time. I maintain that nonstandard logics (the logic of abductive reasoning, for instance) at both conceptual and technical levels, along with some cognitive science results, can help us to change that situation. I would like to stress again that the chapter 7 section on "The Logical Structure of Reasons" is a crucial part of the book, for it is there that I explicitly work out my thesis that we can appeal to scientific thinking for models of practical reasoning.

I would also hope that readers come away from *Morality in a Technological World* with a renewed appreciation for the notion of bad faith, which I believe addresses important issues about personal aspects of moral knowledge. It may seem to be, at best, only tangentially related to science and technology, but I argue that the idea of bad faith is a useful and promising topic to consider alongside the other issues in the book. The bad faith construct asks questions about knowledge at the personal level and about how we manage our individual identities, and it is therefore directly related to the idea of knowledge as duty; and knowledge is in turn directly connected to the theme of personal freedom and responsibility. This concept is a natural ally of my argument for committing oneself to greater ethical awareness: bad faith can be seen as a lack of knowledge, a blind spot in one's knowledge of oneself. When in bad faith, people shrink away from the supposed burden of choice and responsibility as a way to protect themselves, but in so doing, they externalize their responsibilities, relinquish their freedom, and consequently diminish their own dignity. I contend that the moral gap caused by bad faith can be bridged or avoided altogether if we have more adequate knowledge and information at our disposal. Overall, I would say that how we as individuals deal (or fail to deal) with information and knowledge needs much greater attention in scientific research, and that such work will, in turn, contribute to psychological research, which is often an overlooked resource in discussions of ethical concerns.

And so we return to the maxim of "respecting people as things": if we do not know how to "respect people as things" (or, as Sartre would say, if we reduce people to facticity), we do not respect and appreciate many aspects of ourselves, and the bad faith cycle is continued. Unfortunately, there are also external factors working to perpetuate bad faith, for the increasing commodification of our lives tends to further emphasize the "facticity" aspect of human beings.

Acknowledging our "condition" is a form of accepting responsibility – it weakens bad faith and is extraordinarily helpful in improving our freedom and the ownership of our destinies. It also heightens awareness of our technologically induced "cyborg" status and of the positive and negative consequences that can accompany it. Here again is yet another version of "knowledge as duty": in so many ways, in so many spheres, a commitment to knowledge can help to encourage responsibility, demystify technology, rectify globalization, and enhance social equality.

Let us begin.

References

Abbey, E. 1975. *The Monkey Wrench Gang*. Lippincott Williams & Wilkins, Philadelphia.

Adam, A. 2002. The ethical dimension of cyberfeminism. In M. Flanagan and A. Booth, eds., pp. 159–174.

Allenby, B. 2005. Technology at the global scale: integrative cognitivism and Earth systems engineering and management. In M. E. Gorman, R. D. Tweney, D. C. Gooding, and A. P. Kincannon, eds., *Scientific and Technological Thinking*. Erlbaum, Mahwah, NJ, and London, pp. 303–343.

Allwein, G., and Barwise, J., eds. 1996. *Logical Reasoning with Diagrams*. Oxford University Press, New York.

Ames, R. T., and Dissanayake, T., eds. 1996. *Self and Deception: A Cross–Cultural Philosophical Inquiry*. State University of New York Press, Albany, NY.

Amselle, J.-L. 2001. *Branchements: anthropologie de l'universelle des cultures*. Flammarion, Paris.

Anderson, D. R. 1986. The evolution of Peirce's concept of abduction. *Transactions of the Charles S. Peirce Society* 22(2): 145–164.

Anderson, M., Anderson, S. L., and Armen, C., eds. 2005. *Machine Ethics: Papers from the AAAI Fall Symposium*, Technical Report FS–05–06. AAAI Press, Menlo Park, CA.

Arecchi, T. 2003. Chaotic neuron dynamics, synchronization, and feature binding: quantum aspects. *Mind and Matter* 1(1): 15–43.

Baars, B. J. 1997. *In the Theater of Consciousness*. Oxford University Press, New York and Oxford.

Baird Callicott, J. 1998. Do deconstructive ecology and sociobiology undermine Leopold's Land Ethic? *Environmental Ethics* 18(4) (1996): 353–372. Also in M. E. Zimmerman, J. B. Callicott, G. Sessions, K. J. Warren, and J. Clark, eds., pp. 145–164.

Baker, R. 2000. *Sex in the Future: The Reproductive Revolution and How It Will Change Us*. Arcade Publishing, New York.

Baldi, P. 2001. *The Shattered Self*. MIT Press, Cambridge, MA.

Barbour I. 1992. Views of technology. In D. Micah Hester and P. J. Ford, eds., pp. 7–34.

Baumeister, R. F., Zhang, L., and Vohs, K. D. 2004. Gossip as cultural learning. *Review of General Psychology* 8(2): 111–121.

Beauchamp T. L. 1999. Principles and other emerging paradigms in bioethics (1994). In H. J. Robinson, R. M. Berry, and K. McDonnell, eds., pp. 22–27.

Beauchamp, T. L., and Childress, J. 1994. *Principles of Biomedical Ethics*, fourth edition. Oxford University Press, Oxford.

Beckett, D. 2000. Internet technology. In D. Langford, ed., pp. 13–46.

Benn, S. 1984. Privacy, freedom, and respect for persons. In F. D. Shoeman, ed., pp. 223–244.

Ben–Ze'ev, A. 2000. *The Subtlety of Emotions*. MIT Press, Cambridge, MA.

Berry, R. M. 1998. From involuntary sterilization to genetic enhancement: the unsettled legacy of *Buck v. Bell. Notre Dame Journal of Law, Ethics and Public Policy* 12(2): 401–448.

Berry, R. M. 1999. Genetic enhancement in the twenty-first century: three problems in legal imagining. *Law Review* 34(3): 715–735.

Biot, J.-B. 1821. On the magnetism impressed on metals by electricity in motion; read at the public setting of the Academy of Sciences, April 2, 1821. *Quarterly Journal of Science* 11: 281–290.

Blounstein, E. J. 1984. Privacy as an aspect of human dignity. In F. Shoeman, ed., pp. 158–202.

Boutilier, C., and Becher, V. 1995. Abduction as belief revision. *Artificial Intelligence* 77: 43–94.

Breggin, P. 1973. Medical news: does behavior-altering surgery imperil freedom? *Journal of the American Medical Association* 225(8): 916–918.

Brown, J. R. 1991. *The Laboratory of the Mind: Thought Experiments in the Natural Sciences*. Routledge, London and New York.

Buck, S. J. 1998. *The Global Commons: An Introduction*. Earthscan, London.

Burtley, J., ed. 1999. *The Genetic Revolution and Human Rights: The Oxford Amnesty Lectures 1998.* Oxford University Press, Oxford.

Bush, C. G. 2000. Women and the assessment of technology. In M. E. Winston and R. D. Edelbach, eds., pp. 69–82.

Bynum, T. W. 2004. Ethical challenges to citizens of 'The Automatic Age': Norbert Wiener on the Information Society. *Journal of Information, Communication and Ethics in Society* 2: 65–74.

Bynum, T. W. 2005. Norbert Wiener's vision: the impact of the 'automatic age' on our moral lives. In R. Cavalier, ed., *The Impact of the Internet on Our Moral Lives*. State University of New York Press, Albany, NY, pp. 11–25.

Bynum, T. W., and Rogerson, S., eds. 2004. *Computer Ethics and Professional Responsibility*. Blackwell, Malden, MA.

Callon, M. 1994. Four models for the dynamics of science. In S. Jasanoff, G. E. Markle, J. C. Petersen, and T. J. Pinch, eds., *Handbook of Science and Technology Studies*. Sage, Los Angeles, pp. 29–63.

Callon, M. 1997. Society in the making: the study of technology as a tool for sociological analysis. In W. E. Bjiker, T. P. Hughes, and T. Pinch, eds., *The Social Construction of Technological Systems*. MIT Press, Cambridge, MA, pp. 83–106.

Callon, M., and Latour, B. 1992. Don't throw the baby out with the bath school! A reply to Collins and Yearley. In A. Pickering, ed., *Science as Practice and Culture*. University of Chicago Press, Chicago and London, pp. 343–368.

Carlson, W. B., and Gorman, M. E. 1990. Understanding invention as a cognitive process: the case of Alexander Graham Bell, Thomas Edison and the telephone. *Science, Technology and Human Values* 15(2):131–164.

Carse, A. L. 1999. Facing up to moral perils: the virtue of care in bioethics (1996). In H. J. Robinson, R. M. Berry, and K. McDonnell, eds., pp. 51–59.

Carsten Stahl, B. 2004. Information, ethics, and computers: the problem of autonomous moral agents. *Minds and Machines* 14: 67–83.

Chopra, S., and White, L. 2004. Artificial agents: personhood in law and philosophy. In R. López de Mántaras and L. Saitta, eds., *Proceedings of the 16th European Conference on Artificial Intelligence*. IOS Press, Amsterdam, pp. 635–639.

Clark, A. 1997. *Being There: Putting Brain, Body and Word Together Again*. MIT Press, Cambridge, MA.

Clark, A. 2002. Towards a science of the bio-technological mind. *International Journal of Cognition and Technology* 1(1): 21–33.

Clark, A. 2003. *Natural-Born Cyborgs: Minds, Technologies, and the Future of Human Intelligence*. Oxford University Press, Oxford and New York.

Clark J. 1998. Introduction to Part Four: Political Ecology. In M. E. Zimmerman, J. B. Callicott, G. Sessions, K. J. Warren, and J. Clark, eds., pp. 345–363.

Compaine, B. M. 2001. *The Digital Divide: Facing a Crisis or Creating a Myth?* MIT Press, Cambridge, MA.

Coombes, S. 2001. Sartre's concept of bad faith in relation to the Marxist notion of false consciousness: inauthenticity and ideology re-examined. *Cultural Logic* 4(2) at: <http://eserver.org/clogic/>.

Cornuéjols, A., Tiberghien, A., and Collet, G. 2000. A new mechanism for transfer between conceptual domains in scientific discovery and education. In special issue, *Model-Based Reasoning in Scientific Discovery: Learning and Discovery*, L. Magnani, N. J. Nersessian, and P. Thagard, eds. *Foundations of Science* 5(2): 129–155.

Damasio, A. R. 1994. *Descartes' Error*. Putnam, New York.

Damasio, A. R. 1999. *The Feeling of What Happens*. Hartcourt Brace, New York.

Dascal, M. 2002. Language as a cognitive technology. *International Journal of Cognition and Technology* 1(1): 35–61.

Davidson, D. 1970. How is weakness of the will possible? In J. Feinberg, ed., pp. 93–113.

Davidson, D. 1985. Deception and division. In E. W. LePore and B. P. McLaughlin, eds., *Actions and Events: Perspectives on the Philosophy of Donald Davidson*. Blackwell, Oxford.

Davis, N. D., and Splichal, S. L. 2000. *Access Denied: Freedom of Information in the Information Age*. Iowa State University Press, Ames, IA.

Davis, W. H. 1972. *Peirce's Epistemology*. Nijhoff, The Hague.

Davy, H. 1821. On the magnetic phenomena produced by electricity. *Philosophical Transactions* 111: 7–19.

Dawkins, R. 1989. *The Selfish Gene* (1976), second edition. Oxford University Press, Oxford.

Dawkins, R. 1998. What's wrong with cloning. In M. C. Nussbaum and C. R. Sunstein, eds., pp. 54–66.

Dehaene, S., and Naccache, L. 2001. Towards a cognitive neuroscience of consciousness. In S. Dehaene, ed., *The Cognitive Neuroscience of Consciousness*. MIT Press, Cambridge, MA, pp. 1–37.

Delgado, J. M., and Anshen, R. N., eds. 1969. *Physical Control of the Mind: Toward a Psychocivilized Society*. Harper and Row, New York.

DeMarco, J. P. 1994. *A Coherence Theory in Ethics*. Rodopi, Amsterdam.

Dennett, D. 1984. *Elbow Room: The Variety of Free Will Worth Wanting*. MIT Press, Cambridge, MA.

Dennett, D. 1991. *Consciousness Explained*. Little, Brown, Boston.

Dennett, D. 2003. *Freedom Evolves*. Viking, New York.

de–Shalit, A. 1998. Is liberalism environment-friendly? (1995). In M. E. Zimmerman, J. B. Callicott, G. Sessions, K. J. Warren, and J. Clark, eds., pp. 386–406.

de Sousa, R. B. 1970. Self–deception. *Inquiry* 13: 308–321.

de Sousa, R. B. 1988. Emotion and self-deception. In B. P. McLaughlin and A. O. Rorty, eds., pp. 325–341.

de Waal, F. 2006. *Primates and Philosophers: How Morality Evolved*. Princeton University Press, Princeton and Oxford.

Dibbel, J. 1993. A rape in the cyberspace. In D. Micah Hester and P. J. Ford, eds., pp. 439–452.

Dunbar, R. I. M. 2004. Gossip in evolutionary perspective. *Review of General Psychology* 8(2): 100–110.

Dworkin, R. 1993. *Life's Dominion*. Knopf, New York.

Dworkin, R. 2000. *Sovereign Virtue: The Theory and Practice of Equality*. Harvard University Press, Cambridge, MA.

Dyson, F. 2000. Technology and social justice. In M. E. Winston and R. D. Edelbach, eds., pp. 138–148.

Edgar, S. L. 2000. *Morality and Machines: Perspectives in Computer Ethics*. Jones and Bartlett, Sudbury, MA.

Egan, D., and Howell, E. A., eds., 2001. *Historical Ecology Handbook: A Restorationist's Guide to Reference Ecosystems*. Island Press, Washington, DC.

Elgesem, D. 1994. Privacy, respect for persons, and risk. In D. Micah Hester and P. J. Ford, eds., pp. 256–277.

Eskridge, W. N., and Stein, E. 1998. Queer clones. In M. C. Nussbaum and C. R. Sunstein, eds., pp. 95–113.

Evans, D. 2001. *Emotion*. Oxford University Press, New York and Oxford.

Faraday, M. 1821–22. Historical sketch on electromagnetism. *Annals of Philosophy* 18: 195–200, 274–290; 19: 107–121.

Feinberg, J. 1992. *Freedom and Fulfillment*. Princeton University Press, Princeton, NJ.

Feinberg, J., ed. 1970. *Moral Concepts*. Oxford University Press, Oxford.

Finnis, J. 1980. *Natural Law and Natural Rights*. Oxford University Press, Oxford.

Fins, J. J. 2004. Neuromodulation, free will and determinism: lessons from the psychosurgery debate. *Clinical Neuroscience Research* 4: 113–118.

Flach, J. M., and Warren, R. 1995. Active psychophysics: the relation between mind and what matters. In J. M. Flach, J. Hancock, P. Caird, and K. Vincente, eds., *Global Perspectives on the Ecology of Human-Machine Systems*. Erlbaum, Hillsdale, NJ, pp. 189–209.

Flanagan, M., and Booth, A., eds. 2002. *Reload: Rethinking Women + Cyberculture*. MIT Press, Cambridge, MA.

Flavin, C. 2000. Power shock: the next energy revolution. In M. E. Winston and R. D. Edelbach, eds., pp. 275–294.

Fletcher, P. C., Happé, F., Frith, U., Baker, S. C., Donlan, R. J., Frackowiak, R. S. J., and Frith, C. D. 1999. Other minds in the brain: a functional imaging study of theory of mind in story comprehension. *Cognition* 57: 109–128.

Floridi, L. 1999. *Philosophy and Computing*. Routledge, London and New York.

Floridi, L. 2002a. Information ethics: an environmental approach to the digital divide. *Philosophy in the Contemporary World* 9(1): 39–45.

Floridi, L. 2002b. On the intrinsic value of information objects and the infosphere. *Ethics and Information Technology* 4: 287–304.

Floridi, L., ed. 2004. *Blackwell Guide to Philosophy of Computing and Information*. Malden, MA.

Floridi, L., and J. W. Sanders. 2004a. The method of abstraction. In M. Negrotti, ed., *Yearbook of the Artificial. Nature, Culture, and Technology. Models in Contemporary Sciences*. Peter Lang, Bern, pp. 177–220.

Floridi,L., and J. W. Sanders. 2004b. On the morality of artificial agents. *Minds and Machines* 14: 349–379.

Ford, K. M., Glymour, C. and Hayes, J. 1995. *Android Epistemology*. AAAI Press and MIT Press, Cambridge, MA.

Foucault, M. 1979. *Discipline and Punish: The Birth of the Prison* (1975), translated by A. Sheridan. Vintage Books, New York.

Fried, C. 1984. Privacy. In F. Shoeman, ed., pp. 202–222.

Fuller, D. A. 1999. *Sustainable Marketing: Managerial-Ecological Issues*. Sage, Thousand Oaks, CA.

Furedi, F. 2002. *Culture of Fear: Risk-taking and the Morality of Low Expectation*, second revised edition. Continuum, London.

Gabbay, D. M., and Ohlbach, H. J., eds., 1996. *Practical Reasoning: International Conference on Formal and Applied Practical Reasoning, FAPR '96, Bonn, Germany, June 3–7, 1996. Proceedings*. Springer, Berlin.

Gabbay, D. M., and Woods, J. 2005. *The Reach of Abduction*. North-Holland, Amsterdam. (vol. 2 of *A Practical Logic of Cognitive Systems*).

Galison, P. 1997. *Images and Logic: A Material Culture of Microphysics*. University of Chicago Press, Chicago.

Gallese, V. 2006. Intentional attunement: a neurophysiological perspective on social cognition and its disruption in austin. *Brain Research* 1079: 15–24.

Gallese, V., Fadiga, L., Fogassi, L. and Rizzolatti, G. 1966. Action recognition in the premotor cortex. *Brain* 119(2): 593–609.

Gardner, R. L. 1999. Cloning and individuality. In J. Burtley, ed., pp. 29–37.

Gavison, R. Privacy and the limits of law. In F. Shoeman, ed., pp. 346–402.

Gazzaniga, M. 2005. *The Ethical Brain*. Dana Press, New York and Washington.

Gershenfeld, N. 1999. *When Things Start to Think*. Henry Holt, New York.

Gerstein, R. S. 1984. Privacy and self-incrimination. In F. Shoeman, ed., pp. 245–264.

Gibson, J. J. 1979. *The Ecological Approach to Visual Perception*. Houghton Mifflin, Boston.

Giddens, A. 1990. *The Consequences of Modernity*. Stanford University Press, Stanford, CA.

Ginsberg, M. L., ed. 1987. *Readings in Nonmonotonic Reasoning*. Morgan Kaufmann, Los Altos, CA.

Gips, J. 1995. Towards the ethical robot. In K. M. Ford, C. Glymour and J. Hayes, eds., pp. 243–252.

Glassner, B. 1999. *The Culture of Fear: Why Americans Are Afraid of the Wrong Things*. Basic Books, New York.

Godwin, M. 1998. *Cyber Rights: Defending Free Speech in the Digital Age*. Random House, Toronto.

Goldman, A. 1992. *Liaisons: Philosophy Meets the Cognitive and Social Sciences*. MIT Press, Cambridge, MA.

Gooding, D. 1990. *Experiment and the Making of Meaning*. Kluwer, Dordrecht.

Gooding, D. 2004. Cognition, construction, and culture: visual theories in the sciences. *Journal of Cognition and Culture* 4: 551–593. Special issue on cognitive anthropology edited by C. Heinze.

Gooding, D. 2005. Seeing the forest for the trees: visualization, cognition, and scientific inference. In M. E. Gorman, R. D. Tweney, D. C. Gooding, and A. P. Kincannon, eds., pp. 173–217.

Gorman, M. 1997. Mind in the world: cognition and practice in the invention of the telephone. *Social Studies of Science* 27: 583–624.

Gorman, M. 2005. Level of expertise and trading zones: combining cognitive and social approaches to technology studies. In M. E. Gorman, R. D. Tweney, D. C. Gooding, and A. P. Kincannon, eds., pp. 287–302.

Gorman, M. E., Tweney, R. D., Gooding, D. C., and Kincannon, A. P., eds. 2005. *Scientific and Technological Thinking*. Erlbaum, Mahwah, NJ and London.

Gotterbarn, D. 1991. Computer ethics: responsibility regained. *National Forum: The Phi Beta Kappa Journal* 71: 26–31.

Gotterbarn, D. 2000. Virtual information and the software engineering code of ethics. In D. Langford, ed., pp. 200–219.

Gotterbarn, D. 2001. Informatics and professional responsibility. *Science and Engineering Ethics* 7(2): 221–230.

Goudie, A. 2000. *The Human Impact on the Natural Environment*. MIT Press, Cambridge, MA.

Gozzi, R. 2001. Computers and the human identity crisis (1993). In D. Micah Hester and P. J. Ford, eds., pp. 147–152.

Greene, J., and Haidt, J. 2002. How (and where) does moral judgment work? *Trends in Cognitive Science* 6(12): 517–523.

Guarini, M. 1996. Mind, morals, and reasons. In D. M. Gabbay and H. J. Ohlbach, eds., pp. 305–317.

Haas, E. B. 1990. *When Knowledge Is Power: Three Models of Change in International Organization*. University of California Press, Berkeley and Los Angeles.

Haight, M. R. 1980. *A Study on Self-Deception*. HarvesterPress, Sussex.

Hameroff, S. R., Kaszniak, A. W., and Chalmers, D. J., eds. 1999. *Toward a Science of Consciousness III: The Third Tucson Discussions and Debates*. MIT Press, Cambridge, MA.
Hammer, E. 1996. Peircean graphs for propositional logic. In G. Allwein and J. Barwise, eds., pp. 129–147.
Hardin, G. 1968. The tragedy of the commons. *Science* 162:1243–1248.
Hardt, M., and Negri, A. 2001. *Empire*. Harvard University Press, Cambridge, MA.
Harris, J. 1998. *Clones, Genes, and Immortality*. Oxford University Press, Oxford.
Harris, J. 1999. Clones, genes, and human rights. In J. Burtley, ed., pp. 61–94.
Harris, J. 2005. Scientific research is a moral duty. *Journal of Medical Ethics* 31: 242–248.
Harris, J., and Holm, S., eds. 1998. *The Future of Human Reproduction*. Clarendon Press, Oxford.
Harvey, D. 1993. The nature of environment: the dialectics of social and environmental change. In R. Miliband and L. Panich., eds., *Real Problems, False Solutions: Socialist Register 1993*. Merlin Press, London.
Himma, K. E. 2004. The moral significance of the interest in information: reflections on a fundamental right to information. Paper presented at the American Conference on Computing and Philosophy, Pittsburgh, August 4–6.
Holyoak, K. J., and Thagard, P. 1995. *Mental Leaps: Analogy in Creative Thought*. MIT Press, Cambridge, MA.
Holzner, B., and Marx, J. H. 1979. *Knowledge Application: The Knowledge System in Society*. Allyn and Bacon, Boston.
Huff, C. 1995. Unintentional power in the design of computing systems. In T. W. Bynum and S. Rogerson, eds., pp. 98–106.
Hull, D. L., and Ruse, M., eds. 1998. *The Philosophy of Biology*. Oxford University Press, Oxford.
Hutchins, E. 1995. *Cognition in the Wild*. MIT Press, Cambridge, MA.
Iacoboni, M. 2003. Understanding intentions through imitation. In S. H. Johnson Frey, ed., pp. 107–138.
Jaukélévitch, V. 1966. *La mauvaise conscience*. PUF, Paris.
Johns, C. 2005. Reflection on the relationship between technology and caring. *Nursing in Critical Care* 10(3): 150–155.
Johnson, D. G. 1991. Proprietary rights in computer software: individual and policy issues. In T. W. Bynum and S. Rogerson, eds., pp. 285–293.
Johnson, D. G. 1994. *Computer Ethics*, second edition. Prentice Hall, Englewood Cliffs, NJ.
Johnson, D. G. 2000. Democratic values and the Internet. In D. Langford, ed., pp. 181–199.
Johnson, D. G. 2004. Integrating ethics and technology. European Conference on Computing and Philosophy (E–CAP2004_ITALY), Abstract, Pavia, Italy, June 2–5.
Johnson, M. 1993. *Moral Imagination: Implications of Cognitive Science in Ethics*. University of Chicago Press, Chicago.
Johnson-Frey, S. H., ed. 2003. *Taking Action: Cognitive Neuroscience Perspectives on Intentional Acts*. MIT Press, Cambridge, MA.

Johnston, M. 1988. Self-deception and the nature of mind. In B. P. McLaughlin and A. O. Rorty, eds., pp. 63–91.

Jonas, H. 1974. Technology and responsibility: reflections on the new tasks of ethics. In *Philosophical Essays: From Ancient Creed to Technological Man*. Prentice-Hall, Englewood Cliffs, NJ, pp. 3–30.

Jonsen, A. R., and Toulmin, S. 1988. *The Abuse of Casuistry: A History of Moral Reasoning*. University of California Press, Berkeley and Los Angeles.

Josephson, J. R. 1998. Abduction-prediction model of scientific inference reflected in a prototype system for model-based diagnosis. *Philosophica* 61(1): 9–17.

Joyce, R. 2006. *The Evolution of Morality*. MIT Press, Cambridge, MA.

Kant, I. 1929. *Critique of Pure Reason*, translated by N. Kemp Smith (reprint 1998; originally published 1787). Macmillan, London.

Kant, I. 1956. *Critique of Practical Reason*, translated by L. W. Beck (originally published 1788). Bobbs-Merrill, Indianapolis, IN.

Kant, I. 1964. *Groundwork of the Metaphysics of Morals* (reprint of the 1956 edition, edited and translated by H. J. Paton, Hutchinson & Co., Ltd., London, third edition; originally published 1785). Harper & Row, New York.

Kass, L. 1999. The wisdom of repugnance (1997). In H. J. Robinson, R. M. Berry, and K. McDonnell, eds., pp. 105–114.

Kates, R. W. 2000. Sustaining life on earth. In M. E. Winston and R. D. Edelbach, eds., pp. 295–302.

Kirkman, R. 2002. Through the looking glass: environmentalism and the problem of freedom. *The Journal of Value Inquiry* 36(1): 27–41.

Kirlik, A. 1998. The ecological expert: acting to create information to guide action. In *Proceedings of the 1998 Conference on Human Interaction with Complex Systems* (HICS '98). IEEE Press, Piscataway, NJ, pp. 15–27.

Kirlik, A. 2005. Reinventing intelligence for an invented world. In R. J. Sternberg and D. D. Preiss, eds., *Intelligence and Technology: The Impact of Tools on the Nature and Development of Human Abilities*. Erlbaum, Mahwah, NJ, pp. 105–134.

Kirsh, D. 1995. The intelligent use of space. *Artificial Intelligence* 73: 31–68.

Kirsh, D. 2006. Distributed cognition: a methodological note. *Pragmatics & Cognition* 14(2): 249–262.

Kitcher, P. 2001. *Science, Truth, and Democracy*. Oxford University Press, Oxford.

Knorr Cetina, K. 1999. *Epistemic Cultures: How the Sciences Make Knowledge*. Harvard University Press, Cambridge, MA.

Kolata, G. 1998. *Clone: The Road to Dolly, and the Path Ahead*. William Morrow, New York.

Kornhuber, H. H., and Deecke, L. 1965. Hirnpotentialänderungen bei Willkürbewegungen un passiven Bewegungen des Menschen: Bereitschaftspotential und reafferente Potentiale. *Pflügers Arch. Ges. Physiol.* 284: 1–17.

Kosslyn, S. M., and Koenig, O. 1992. *Wet Mind: The New Cognitive Neuroscience*. Free Press, New York.

Krauthammer, C. 1999. Creation of headless humans should be banned. In Rantala and Milgram, eds., pp. 81–83.

Kuhn, T. S. 1970. *The Structure of Scientific Revolutions* (1962), second edition. University of Chicago Press, Chicago.

Lackey, D. P. 1989. *The Ethics of War and Peace*. Prentice-Hall, Englewood Cliffs, NJ.

Langford, D., ed. 2000. *Internet Ethics*. St. Martin Press, New York.

Lanzola, G., Stefanelli, M., Barosi, G., and Magnani, L. 1990. NEOANEMIA: a knowledge-based system emulating diagnostic reasoning. *Computer and Biomedical Research* 23: 560–582.

Latour, B. 1987. *Science in Action: How to Follow Scientists and Engineers through Society*. Harvard University Press, Cambridge. MA.

Latour, B. 1988. *The Pasteurization of France*. Harvard University Press, Cambridge, MA.

Latour, B. 1999. *Pandora's Hope*. Harvard University Press, Cambridge, MA.

Lauritzen, P., ed. 2001. *Cloning and the Future of Embryo Research*. Oxford University Press, Oxford and New York.

Law, J. 1993. *Modernity, Myth, and Materialism*. Blackwell, Oxford.

Lazar, A. 1999. Deceiving oneself or self-deceived? On the formation of beliefs 'under the influence'. *Mind* 108: 265–290.

Lederman, S. J., and Klatzky, R. 1990. Haptic exploration and object representation. In M. A. Goodale, ed., *Vision and Action: The Control of Grasping*. Ablex, Norwood, NJ, pp. 98–109.

Leiser, B. M. 2003. Is homosexuality unnatural? In J. Rachels, ed., pp. 144–153.

Leopold, A. 1933. The conservation ethic. *Journal of Forestry* 31: 634–643. Reprinted in the same journal 87(6) (1989): 26–28, 42–45.

Leopold, A. 1998. The land ethic (1949). In his *A Sand County Almanac: and Sketches Here and There*. Oxford University Press, Oxford, 1966. Also in M. E. Zimmerman, J. B. Callicott, G. Sessions, K. J. Warren, and J. Clark, eds., pp. 87–100.

Lewontin, R. 1997. Confusion about cloning. *New York Review of Books* 23: 18–23.

Libet, B. 1999. Do we have free will? In B. Libet, A. Freeman, and K. Sutherland, eds., pp. 45–55.

Libet, B., Freeman, A., and Sutherland, K., eds. 1999. *The Volitional Brain: Towards a Neuroscience of Free Will*. Imprint Academic, Thorverton, UK.

Libet, B., Gleason, C. A., Wright, E. W., and Pearl, D. K. 1983. Time of conscious intention to act in relation to onset of cerebral activity (readiness-potential): the unconscious initiation of a freely voluntary act. *Brain* 106: 623–642.

Loye, D. 2002. The moral brain. *Brain and Mind* 3: 133–150.

Łukasiewicz, J. 1970. Creative elements in science (1912). In his *Selected Works*. North Holland, Amsterdam, pp. 12–44.

Lukaszewicz, W. 1990. *Non-Monotonic Reasoning: Formalization of Commonsense Reasoning*. Horwood Publishing, Chichester, UK.

Lycan, W. 1988. *Judgment and Justification*. Cambridge University Press, Cambridge.

MacLean, P. D. 1990. *The Triune Brain in Evolution Role in Paleocerebral Functions*. Plenum, New York.

Magnani, L. 1988. Epistémologie de l'invention scientifique. *Communication & Cognition* 21: 273–291.

Magnani, L. 1991. *Epistemologia applicata*. Marcos y Marcos, Milan.

Magnani, L. 1997. *Ingegnerie della conoscenza. Introduzione alla filosofia computazionale*. Marcos y Marcos, Milan.

Magnani, L. 1999. Model-based creative abduction. In L. Magnani, N.J. Nersessian, and P. Thagard, eds., pp. 219–238.

Magnani, L. 2000. Action-based abduction in science. In A. Aliseda and D. Pearce, eds., Workshop Scientific Reasoning in Artificial Intelligence and Philosophy of Science, *ECAI 2000 Workshop Notes*, Berlin, pp. 46–51.

Magnani, L. 2001a. *Abduction, Reason, and Science: Processes of Discovery and Explanation*. Kluwer Academic/Plenum Publishers, New York.

Magnani, L. 2001b. *Philosophy and Geometry: Theoretical and Historical Issues*. Kluwer Academic, Dordrecht.

Magnani, L. 2001c. Medici automatici. *Kéiron* 9: 40–51.

Magnani, L. 2002. Epistemic mediators and model-based discovery in science. In L. Magnani and N.J. Nersessian, eds., pp. 305–329.

Magnani, L. 2005. Abduction and cognition in human and logical agents. In S. Artemov, H. Barringer, A. Garcez, L. Lamp, and J. Woods, eds., *We Will Show Them: Essays in Honour of Dov Gabbay*, vol. II. College Publications, London. pp. 225–258.

Magnani, L. 2006a. Multimodal abduction: external semiotic anchors and hybrid representations. *Logic Journal of the IGPS* 14(1): 107–136.

Magnani, L. 2006b. Disembodying minds, externalizing minds: how brains make up creative scientific reasoning. In L. Magnani, *Model-Based Reasoning in Science and Engineering: Cognitive Science, Epistemology, Logic*. College Publications, London, pp. 185–202.

Magnani, L. 2006c. Mimetic minds: meaning formation through epistemic mediators and external representations. In A. Loula, R. Gudwin, and J. Queiroz, eds., *Artifical Cognition Systems*. Idea Group Inc., Hershey, PA, pp. 327–357.

Magnani, L. 2007a. Semiotic brains and artificial minds: how brains make up material cognitive systems. In R. Gudwin and J. Queiroz, eds., *Semiotics and Intelligent Systems Development*. Idea Group Inc., Hershey, PA, pp. 1–41.

Magnani, L. 2007b. Animal abduction: from mindless organisms to artifactual mediators. In L. Magnani and Poli, eds., *Model-Based Reasoning in Science, Technology, and Medicine*. Springer, Berlin, forthcoming.

Magnani, L., and Bardone, E. 2005. Abduction and WEB interfaces design. In C. Gahoui, ed., *Encyclopedia of Human Computer Interaction*. Idea Group Inc., Hershey, PA, pp. 1–7.

Magnani, L., and Bardone, E. 2007. Sharing representations and creating changes through cognitive niche construction: the role of affordances. In Y. S. Iwata Oshawa, S. Tsumoto, N. Zhong, Y. Shi, and L. Magnani, eds., *Communications and Discoveries from Multidisciplinary Data*. Berlin, Springer, forthcoming.

Magnani, L., Civita, S., and Previde Massara, G. 1994. Visual cognition and cognitive modeling. In V. Cantoni, ed., *Human and Machine Vision: Analogies and Divergences*. Plenum, New York, pp. 229–243.

Magnani, L., and Dossena, R. 2005. Perceiving the infinite and the infinitesimal world: unveiling and optical diagrams and the construction of mathematical concepts. *Foundations of Science* 10: 7–23.

Magnani, L., and Gennari, R. 1997. *Manuale di logica. Logica classica e del senso comune*. Guerini, Milan.

Magnani, L., and Nersessian, N.J., eds. 2002. *Model-Based Reasoning: Scientific Discovery, Technology, Values*. Kluwer Academic/Plenum Publishers, New York.

Magnani, L., Nersessian, N.J., and Pizzi, C., eds. 2002. *Logical and Computational Aspects of Model-Based Reasoning*. Kluwer Academic Publishers, Dordrecht.

Magnani, L., Nersessian, N.J., and Thagard, P., eds. 1999. *Model-Based Reasoning in Scientific Discovery*. Kluwer Academic/Plenum Publishers, New York.

Magnani, L., Piazza, M., and Dossena, R. 2002. Epistemic mediators and chance morphodynamics: 2nd Workshop on Chance Discovery (CDWS2). In *Proceedings of PRICAI-02 Conference, Workshop on Chance Discovery*, pp. 38–46.

Maguire, G.Q., and McGee, E.M. 2001. Implantable brain chips? Time for debate (1999). In D. Micah Hester and P.J. Ford, eds., pp. 129–141.

Maienschein, J., and Ruse, M. 1999. *Biology and the Foundations of Ethics*. Cambridge University Press, Cambridge.

Maner, W. 1980. *Starter Kit in Computer Ethics*. Helvetia Press, New York (published in cooperation with the National Information and Resource Center for Teaching Philosophy).

Manes, C. 1998. Ecotage. In M.E. Zimmerman, J.B. Callicott, G. Sessions, K.J. Warren, and J. Clark, eds., pp. 457–463.

Marcus, S.J., ed. 2004. *Neuroethics: Mapping the Field*. The Dana Press, New York.

Marquis, D. 2003. Why is abortion immoral? In J. Rachels, ed., pp. 107–113.

Marx, C. 1973. *Grundrisse: Foundations of the Critique of Political Economy (1859)*. translated with a Foreword by M. Nicolaus. Random House, New York.

Matthews, G., Zeidner, M., and Roberts, R.D., eds. 2002. *Emotional Intelligence: Science and Myth*. MIT Press, Cambridge, MA.

Mawhood, J., and Tysver, D. 2000. Law and the internet. In D. Langford, ed., pp. 96–126.

May, L., Friedman, M., and Clark, A., eds. 1996. *Mind and Morals: Essays on Ethics and Cognitive Science*. MIT Press, Cambridge, MA.

McGinn, R. 2000. Technology, demography, and the anachronism of traditional rights. In M.E. Winston and R.D. Edelbach, eds., pp. 125–138.

McHugh, P. 2004. Zygote and clonote: the ethical use of embryonic stem cells. *New England Journal of Medicine* 351: 209–211.

McLaughlin, B.P. 1996. On the very possibility of self-deception. In R.T. Ames and W. Dissanayake, eds., pp. 11–25.

McLaughlin, B.P., and Rorty, A.O., eds. 1988. *Perspectives on Self-Deception*. University of California Press, Berkeley.

Mele, A.R. 2001. *Self-Deception Unmasked*. Princeton University Press, Princeton, NJ.

Micah Hester, D., and Ford, P.J., eds. 2001. *Computers and Ethics in the Cyberage*. Prentice-Hall, Upper Saddle River, NJ.

Mill, J.S. 1966. *On Liberty* (1859). In J.S. Mill, *On Liberty, Representative Government, The Subjection of Women*, twelfth edition. Oxford University Press, Oxford, pp. 5–141.

Millgram, E. 2001a. The current state of play. In E. Millgram, ed., 2001. pp. 1–26.

Millgram, E., ed. 2001b. *Varieties of Practical Reasoning*. MIT Press, Cambridge, MA.

Minski, M. 1985. *The Society of Mind*. Simon and Schuster, New York.

Mithen, S. 1996. *The Prehistory of the Mind: A Search for the Origins of Art, Religion, and Science*. Thames and Hudson, London.

Modell, A. H. 2003. *Imagination and the Meaningful Brain*. MIT Press, Cambridge, MA.

Mohr, R. D. 2003. Gay basics: some questions, facts, and values. In J. Rachels, ed., pp. 128–143.

Moll, J., de Oliveira-Souza, R., Eslinger, P. J., Bramati, I. E., Mourão-Miranda, J., Andreiuolo, P. D., and Pessoa L. 2002. The neural correlates of moral sensitivity: a functional magnetic resonance imaging investigation of basic and moral emotions. *The Journal of Neuroscience* 22(7): 2730–2736.

Moll, J., Zahn, R., de Oliveira-Souza, R., Krueger, F., and Grafman, J. 2005. The neural basis of human moral cognition. *Nature Reviews Neuroscience* 6: 799–809.

Moor, J. H. 1985. What is computer ethics? *Metaphilosophy* 16(4): 266–275.

Moor, J. H. 1997. Towards a theory of privacy in the information age. *Computers and Society* 27: 27–32.

Moor, J. H. 2005. The nature and importance of Machine Ethics. In M. Anderson, S. L. Anderson, and C. Armen, eds., p. 78.

Moor, J. H., and Bynum, T. W., eds. 2002. *Cyberphilosophy*. Blackwell, Malden, MA.

Moore, S. C., and Oaksford, M., eds. 2002. *Emotional Cognition: From Brain to Behaviour*. John Benjamins, Amsterdam.

Morgan, M. S., and Morrison, M., eds. 1999. *Models as Mediators: Perspectives on Natural and Social Science*. Cambridge University Press, Cambridge.

Mumford, L. 1961. Assimilation of the machine: new cultural values. In D. Micah Hester and P. J. Ford, eds., pp. 3–7.

Mun Chan, H., and Gorayska, B. 2002. Critique of pure technology. *International Journal of Cognition and Technology* 1(1): 63–84.

Naess, A. 1998. The deep ecological movement: some philosophical aspects. In M. E. Zimmerman, J. B. Callicott, G. Sessions, K. J. Warren, and J. Clark, eds., pp. 193–211.

Nagel, T. 1979. *Mortal Questions*. Cambridge University Press, Cambridge.

Nagle, J. C. 1998a. Endangered species wannabees. *Seton Hall Law Review* 29: 235–255.

Nagle, J. C. 1998b. Playing Noah. *Minnesota Law Review* 82: 1171–1260.

Nersessian, N. J. 1995a. Opening the black box: cognitive science and history of science. Technical Report GIT-COGSCI 94/23, July. Cognitive Science Report Series, Georgia Institute of Technology, Atlanta, GA. Partially published in *Osiris* 10: 194–211.

Nersessian, N. J. 1995b. Should physicists preach what they practice? Constructive modeling in doing and learning physics. *Science and Education* 4: 203–226.

Nersessian, N. J. 1999. Model-based reasoning in conceptual change. In L. Magnani, N. J. Nersessian, and P. Thagard, eds., pp. 5–22.

Norman, D. A. 1993. *Things That Make Us Smart: Defending Human Attributes in the Age of the Machine*. Addison-Wesley, Reading, MA.

Norman, D.A. 1999. *The Invisible Computer*. MIT Press, Cambridge, MA.

Norton, B.G. 2002. Building demand models to improve environmental policy process. In L. Magnani and N.J. Nersessian, eds., *Model-Based Reasoning: Science, Technology, Values*. Kluwer Academic/Plenum Publishers, New York, pp. 191–208.

Nussbaum, M.C. 2001. *Upheavals of Thought: The Intelligence of Emotions*. Cambridge University Press, Cambridge.

Nussbaum, M.C., and Sunstein, C.R., eds. 1998. *Clones and Clones: Facts and Fantasies about Human Cloning*. W.W. Norton, New York.

Oatley, K. 1991. *Morality and the Emotions*. Routledge, London.

Oatley, K. 1992. *Best Laid Schemes: The Psychology of Emotions*. Cambridge University Press, Cambridge.

Oatley, K. 1996. Inference in narrative and science. In D.R. Olson and N. Torrance, eds., pp. 123–140.

Oatley, K., and Johnson-Laird, P.N. 1987. Towards a cognitive theory of emotions. *Cognition and Emotions* 1: 29–50.

O'Connor, J. 1998. Socialism and ecology. In M.E. Zimmerman, J.B. Callicott, G. Sessions, K.J. Warren, and J. Clark, eds., pp. 407–415.

Okada, T., and Simon, H.A. 1997. Collaborative discovery in a scientific domain. *Cognitive Science* 21: 109–146.

O'Neill, O. 2001. Consistency in action. In E. Millgram, ed., pp. 301–328.

Ostrom, E. 1990. *Governing the Commons: The Evolution of Institutions for Collective Action*. Cambridge University Press, Cambridge.

Paluch, S. 1967. Self-deception. *Inquiry* 10: 268–278.

Parsons, S. 2001. *Qualitative Methods for Reasoning under Uncertainty*. MIT Press, Cambridge, MA.

Pascal, B. 1967. *The Provincial Letters* (1657), translated with an introduction by A.J. Krailsheimer. Penguin Books, Baltimore.

Passingham, R.E. 1993. *The Frontal Lobes and Voluntary Action*. Oxford University Press, Oxford.

Pears, D. 1986. The goals and strategies of self-deception. In J. Elster, ed., *The Multiple Self*. Cambridge University Press, Cambridge, pp. 59–78.

Pech, T., and Padis, M.-O. 2004. *Multinationales du coeur: la politique des ONG*. Seuil, Paris.

Peirce, C.S. 1955. Abduction and induction. In *Philosophical Writings of Peirce*, ed. J. Buchler. Dover, New York, pp. 150–156.

Peirce, C.S. 1931–58 (*CP*). *Collected Papers*, 8 vols., ed. C. Hartshorne and P. Weiss (vols. I–VI) and A.W. Burks (vols. VII–VIII). Harvard University Press, Cambridge, MA.

Pellegrino, E.D. 1999. Toward a virtue-based normative ethics for the health professions (1995). In H.J. Robinson, R.M. Berry, and K. McDonnell, eds., pp. 43–50.

Perkins, D. 2003. *King Arthur's Round Table: How Collaborative Conversations Create Smart Organizations*. Wiley, Chichester.

Piaget, J. 1974. *Adaptation and Intelligence*. University of Chicago Press, Chicago.

Picard, R.W. 1997. *Affective Computing*. MIT Press, Cambridge, MA.

Pickering, A. 1995. *The Mangle of Practice: Time, Agency, and Science*. University of Chicago Press, Chicago and London.

Pirandello, L. 1990. *One, No One, and One Hundred Thousand* (1926), translated with an introduction by W. Weaver. Eridanos Press, Boston.

Plumwood, V. 1998. Nature, self, and gender: feminism, environmental philosophy, and the critique of rationalism. *Hypatia* 6(1) (1991): 3–27. Also in M. E. Zimmerman, J. B. Callicott, G. Sessions, K. J. Warren, and J. Clark, eds., pp. 291–314.

Posner, E.A., and Posner, R.A. 1998. The demand for human cloning. In M. C. Nussbaum and C. R. Sunstein, eds., pp. 233–261.

Posner, M. I. 1994. Attention: the mechanism of consciousness. *Proc. National Acad. of Science, U.S.A.* 91(16): 7398–7402.

Posner, R. 1981. The economic analysis of law. In his *The Economics of Justice*. Cambridge University Press, Cambridge, pp. 29–33.

Postman, N. 2000. Invisible technologies. In M. E. Winston and R. D. Edelbach, eds., pp. 81–92.

Powers, T. M. 2004. Intentionality and moral agency in computers. European Conference on Computing and Philosophy (E–CAP2004_ITALY), Abstract, Pavia, Italy, June 2–5.

Powers, W. T. 1973. *Behavior: The Control of Perception*. Aldine, Chicago.

Putnam, H. 1999. Cloning people. In J. Burtley, ed., pp. 1–13.

Rachels, J. 1985. *The End of Life: Euthanasia and Morality*. Oxford University Press, Oxford.

Rachels, J. 1999. *The Elements of Moral Philosophy*. McGraw-Hill College, Boston.

Rachels, J., ed. 2003. *The Right Thing to Do*. McGraw-Hill College, Boston.

Radin, M. J. 1999. Market-inalienability (1987). In H. J. Robinson, R. M. Berry, and K. McDonnell, eds., pp. 114–129.

Rajan, R. G., and Zingales, L. 2003. *Saving Capitalism from the Capitalists: Unleashing the Power of Financial Markets to Create Wealth and Spread Opportunity*. Crown Business, New York.

Ramoni, M., Magnani, L., and Stefanelli, M. 1989. Una teoria formale del ragionamento diagnostico. In *Atti del Primo Convegno della Associazione Italiana per l'Intelligenza Artificiale*. Cenfor, Genoa, pp. 267–283.

Ramoni, M., Stefanelli, M., Magnani, L., and Barosi, G. 1992. An epistemological framework for medical knowledge-based systems. *IEEE Transactions on Systems, Man, and Cybernetics* 22(6): 1361–1375.

Rantala, M. L., and Milgram, A. J., eds. 1999. *Cloning: For and Against*. Open Court, Chicago and La Salle, IL.

Rao, R. 1999. Assisted reproductive technology and the threat to the traditional family (1996). In H. J. Robinson, R. M. Berry, and K. McDonnell, eds., pp. 91–96.

Rawls, J. 1971. *A Theory of Justice*. Harvard University Press, Cambridge, MA.

Regan, T. 1998. Animal rights. *Environmental Ethics* 2(2) (1980): 99–120. Also in M. E. Zimmerman, J. B. Callicott, G. Sessions, K. J. Warren, and J. Clark, eds., pp. 41–55.

Reiman, J. H. 1984. Privacy, intimacy, personhood. In F. Shoeman, ed., pp. 300–316.

Reiman, J. H. 1995. Driving to the Panopticon: a philosophical exploration of the risks to privacy posed by the highway technology of the future. *Computer and High Technology Law Journal* 11(1): 27–44.

Resnik, D. B. 1988. *The Ethics of Science: An Introduction*. Routledge, London and New York.

Rheingold, H. 1991. *Virtual Reality*. Seiter and Warburg, London.

Rifkin, J. 2000. The end of work. In M. E. Winston and R. D. Edelbach, eds., pp. 164–171.

Rizzolatti, G., and Arbib, M. A. 1988. Language within our grasp. *Trends in Neuroscience* 21(5): 188–194.

Rizzolatti, G., Carmada, R., Gentilucci, M., Luppino, G., and Matelli, M. 1988. Functional organization of Area 6 in the macaque monkey. II Area F5 and control of distal movements. *Experimental Brain Research* 71: 491–507.

Robertson, J. A. 1998. Liberty, identity, and human cloning. *Texas Law Review* 76: 1371–1456.

Robertson, J. A. 1999a. Two models of human cloning. *Hofstra Law Review* 27: 609–638.

Robertson, J. A. 1999b. The presumptive primacy of procreative liberty (1994). In H. J. Robinson, R. M. Berry, and K. McDonnell, eds., pp. 74–86.

Robinson, H. J., Berry, R. M., and McDonnell, K., eds. 1999. *A Health Law Reader: An Interdisciplinary Approach*. Carolina Academic Press, Durham, NC.

Rogerson, S. 1996. The ethics of computing: the first and second generations. *The UK Business Ethics Network News* (Spring): 1–4.

Rorty, A. O. 1972. Belief and self-deception. *Inquiry* 15: 387–410.

Rorty, A. O. 1988. The deceptive self: liars, layers, and lairs. In B. P. McLaughlin and A. O. Rorty, eds., pp. 11–28.

Rorty, A. O. 1996. User-friendly self-deception: a traveler's manual. In R. T. Ames and W. Dissanayake, eds., pp. 387–410.

Rosenau, J. N. 1992. Governance, order and change in world politics. In J. N. Rosenau and E.-O. Czempiel, eds., *Governance without Government: Order and Change in World Politics*. Cambridge University Press, Cambridge, pp. 1–29.

Rossetti, Y., and Pisella, L. 2003. Mediate responses as direct evidence for intention: neuropsychology of not-to, not-how, and not-there tasks. In S. H. Johnson Frey, ed., pp. 67–105.

Rowlands, M. 1999. *The Body in Mind*. Cambridge University Press, Cambridge.

Russo, E., and Cove, D. 2000. Frankenstein's monster and other horror stories. In M. E. Winston and R. D. Edelbach, eds., pp. 241–246.

Sachs, J. 2000. International economics: unlocking the mysteries of globalization. In M. E. Winston and R. D. Edelbach, eds., pp. 181–188.

Safire, W. 2005. The but-what-if factor. *New York Times*, May 16.

Sahdra, B., and Thagard, P. 2003. Self-deception and emotional coherence. *Minds and Machines* 13: 213–231.

Sartre, J. P. 1956. *Being and Nothingness: An Essay on Phenomenological Ontology (1943)*, translated and with an introduction by H. E. Barnes. Philosophical Library, New York.

Sayre-McCord, G. 1996. Coherentist epistemology and moral theory. In W. Sinnott-Armstrong and M. Timmons, eds., pp. 137–189.

Schaffner, K. 2002. Neuroethics: reductionism, emergence, and decision-making capacities. In S. J. Marcus, ed., pp. 27–33.

Schiffer, S. 1976. A paradox of desire. *American Philosophical Quarterly* 13(3): 195–212.

Schiller, H. 2000. The global information highway. In M. E. Winston and R. D. Edelbach, eds., pp. 171–181.

Searle, J. R. 2001. *Rationality in Action*. MIT Press, Cambridge, MA.

Seebauer, E. G., and Barry, R. L. 2001. *Fundamentals of Ethics for Scientists and Engineers*. Oxford University Press, New York and Oxford.

Shelley, C. 1996. Visual abductive reasoning in archaeology. *Philosophy of Science* 63(2): 278–301.

Shelley, C. 2006. Analogical reasoning with animal models in biochemical research. In L. Magnani, ed., *Model-Based Reasoning in Science and Engineering: Cognitive Science, Epistemology, Logic*. College Publications, London, pp. 203–213.

Shimojima, A. 2002. A logical analysis of graphical consistency proof. In L. Magnani, N. J. Nersessian, and C. Pizzi, eds., pp. 93–116.

Shoeman, F., ed. 1984. *Philosophical Dimensions of Privacy: An Anthology*. Cambridge University Press, Cambridge.

Singer, P. 1998. All animals are equal. *Philosophic Exchange* 1(5), (1974): 243–257. Also in M. E. Zimmerman, J. B. Callicott, G. Sessions, K. J. Warren, and J. Clark, eds., pp. 26–40.

Sinnott-Armstrong, W. 1996. Moral skepticism and justification. In W. Sinnott-Armstrong and M. Timmons, eds., pp. 3–48.

Sinnott-Armstrong, W., and Timmons, M., eds. 1996. *Moral Knowledge? New Readings in Moral Epistemology*. Oxford University Press, Oxford.

Sloman, A. A., and Lagnado, D. A. 2005. Do we "do"? *Cognitive Science* 29: 5–39.

Sober, E. 1998. What is evolutionary altruism? In D. L. Hull and M. Ruse, eds., pp. 459–478.

Spinello, R. A. 2000. Information integrity. In D. Langford, ed., pp. 158–180.

Stallman, R. 1991. Why software should be free. In T. W. Bynum and S. Rogerson, eds., pp. 294–310.

Stapp, H. 2001. Quantum theory and the role of mind in nature. *Foundations of Physics* 31(10): 1468–1499.

Stefanelli, M., and Ramoni, M. 1992. Epistemological constraints on medical knowledge-based systems. In D. A. Evans and V. L. Patel, eds., *Advanced Models of Cognition for Medical Training and Practice*. Springer, Berlin, pp. 3–20.

Steinbock, B. 1992. *Life before Birth: The Moral and Legal Status of Embryos and Fetuses*. Oxford University Press, New York.

Steinbock, B. 2001. Respect for human embryos. In P. Lauritzen, ed., pp. 21–33.

Steinhart, E. 1999. Emergent values for automations: ethical problems of life in the generalized internet. *Journal of Ethics and Information Technology* 1(2): 155–160.

Strong, D. 2000. Technological subversion. In M. E. Winston and R. D. Edelbach, eds., pp. 148–159.

Svevo, I. 1958. *Confessions of Zeno* (1923), translated by B. De Zoete. Vintage Books, New York (reissue edition, 1989).

Swoboda, N., and Allwein, G. 2002. A case study of the design and implementation of heterogeneous reasoning systems, In L. Magnani, N.J. Nersessian, and C. Pizzi, eds., pp. 3–20.

Talbott, W.J. 1995. Intentional self-deception in a single-coherent self. *Philosophy and Phenomenological Research* 55(1): 27–74.

Tavani, H. 2000. Privacy and security. In D. Langford, ed., pp. 65–95.

Tavani, H. 2002. The uniqueness debate in computer ethics: what exactly is at issue, and why does it matter? *Ethics and Information Technology* 4: 37–54.

Teeple, G. 2000. *Globalization and the Decline of Social Reform*. Garamond Press, Aurora, Ontario.

Thagard, P. 1988. *Computational Philosophy of Science*. MIT Press, Cambridge, MA.

Thagard, P. 1989. Explanatory coherence. *Behavioral and Brain Sciences* 12(3): 435–467.

Thagard, P. 1992. *Conceptual Revolutions*. Princeton University Press, Princeton, NJ.

Thagard, P. 1997. Collaborative knowledge. *Noûs* 31: 242–261.

Thagard, P. 2000. *Coherence in Thought and in Action*. MIT Press, Cambridge, MA.

Thagard, P. 2001. How to make decisions: coherence, emotion, and practical inference. In E. Millgram, ed., pp. 355–371.

Thagard, P., and Verbeurgt, K. 1998. Coherence as constraint satisfaction. *Cognitive Science* 22: 1–24.

Thomas, N.J. 1999. Are theories of imagery theories of imagination? An active perception approach to conscious mental content. *Cognitive Science* 23(2): 207–245.

Thomson, A. 1999. *Critical Reasoning in Ethics*. Routledge, London.

Tribe, L.H. 2000. The constitution in cyberspace. In M.E. Winston and R.D. Edelbach, eds., pp. 223–231.

Urban Walker, M. 1996. Feminist skepticism, authority and transparency. In W. Sinnott-Armstrong and M. Timmons, eds., pp. 267–292.

van den Hoven, J. 2000. The Internet and varieties of moral wrongdoing. In D. Langford, ed., pp. 127–157.

van Wel, L., and Royakkers, L. 2004. Ethical issues in web data mining. *Ethics and Information Technology* 6: 129–140.

Velmans, M. 2000. *Understanding Consciousness*. Routledge, London.

Vico, G. 1968. *The New Science of Giambattista Vico*, revised translation of the third edition (1744) by T.G. Bergin and M.H. Fisch. Cornell University Press, Ithaca, NY.

Vitiello, G. 2001. *My Double Unveiled: The Dissipative Quantum Model of the Brain*. John Benjamins, Amsterdam.

Vogler, J. 2000. *The Global Commons: Environmental and Technological Governance*. Wiley, Chichester, UK.

Von Krogh, G., Ichijo, K., and Nonaka, I. 2000. *Enabling Knowledge Creation: How to Unlock the Mistery of Tacit Knowledge and Release the Power of Innovation*. Oxford University Press, Oxford.

Wachbroit, R. 2000. Genetic encores: the ethics of human cloning. In M. E. Winston and R. D. Edelbach, eds., pp. 253–259.

Wagar, B. M., and Thagard, P. 2004. Spiking Phineas Gage: a neurocomputational theory of cognitive-affective integration in decision making. *Psychological Review* 111(1): 67–79.

Wajcman, J. 2000. Reproductive technology: delivered into men's hands. In M. E. Winston and R. D. Edelbach, eds., pp. 259–275.

Warwick, K. 2003. Cyborg morals, cyborg values, cyborg ethics. *Ethics and Information Technology* 5: 131–137.

Weckert, J. 2000. What is new or unique about Internet activities? In D. Langford, ed., pp. 47–64.

Wegner, D. M. 2002. *The Illusion of Conscious Will*. MIT Press, Cambridge, MA.

Weiser, M. 1991. The computer for the twenty-first century. *Scientific American* (September): 99–110.

Weiskrantz, L. 1997. *Consciousness Lost and Found: A Neuropsychological Exploration*. Oxford University Press, New York and Oxford.

Wel, van L., and Royakkers, L. 2004. Ethical issues in web data mining. *Ethics and Information Technology* 6: 129–140.

Wilson, E. O. 1998a. *Consilience: The Unity of Knowledge*. Knopf, New York.

Wilson, E. O. 1998b. On the relationship between evolutionary and psychological definitions of altruism and selfishness. In D. L. Hull and M. Ruse, eds., pp. 479–487.

Wilson, J. Q. 1993. *The Moral Sense*. Free Press, New York.

Winston, M. E., and Edelbach, R. D. 2000. *Society, Ethics, and Technology*. Wadsworth/Thomson Learning, Belmont, MA.

Yerkovich, S. 1977. Gossiping as a way of speaking. *Journal of Communication* 27: 192–196.

Zhang, J. 1997. The nature of external representations in problem solving. *Cognitive Science* 21(2): 179–217.

Zhang, J., and Norman, D. A. 1994. Representations in distributed cognitive tasks. *Cognitive Science* 18: 87–122.

Zimmerman, M. E., Callicott, J. B., Sessions, G., Warren, K. J., and Clark, J., eds., 1998. *Environmental Philosophy: From Animal Rights to Radical Ecology*, Prentice-Hall, Upper Saddle River, NJ.

Index

Lightning Source UK Ltd.
Milton Keynes UK
05 November 2009

145862UK00001B/87/P

9 780521 121798